VALUING PLACE AND PURPOSE

APPLYING GIS

VALUING PLACE AND PURPOSE

GIS FOR LAND ADMINISTRATION

Edited by
Brent Jones
Keith Mann

Esri Press
REDLANDS | CALIFORNIA

Esri Press, 380 New York Street, Redlands, California 92373-8100

Printed in the United States of America.

ISBN: 9781589487062
Library of Congress Control Number: 2021951779

For purchasing and distribution options (both domestic and international), please visit esripress.esri.com.

On the cover: Photograph by Curioso Photography.

CONTENTS

INTRODUCTION

LAND IS AN ESSENTIAL RESOURCE THAT SUPPORTS NEARLY every aspect of a nation's social fabric and economic well-being. The finite supply of land must be governed to support sustainable economic growth, social equity, and access to services, and to prevent conflict. Good land governance controls and manages the effective use of physical space through a combination of policies, processes, and institutions. It includes determining, recording, and disseminating information about the tenure, value, and use of land.

As countries increasingly engage in the global economy, rapid economic, technological, and social transformations also increase the demand for modern land administration systems that support the collection, management, and dissemination of complete, reliable, and authoritative land administration data. Overcoming the lack of available, standardized, high-quality, and usable land data is a significant challenge for a successful economy.

A standards-based land administration solution, grounded in a legal and institutional framework, is critical to document and manage municipal, provincial, state, national, and private-sector land-based resources. It also must support the standard functions of a modern government, such as providing e-government services, property valuation and taxation, access to utilities, and urban and regional planning activities.

Geographic information system (GIS) technology is an essential part of successful land administration and cadastral agencies. GIS provides a comprehensive platform and strategy that allows land administration organizations to modernize and meet new challenges, such as creating efficient, well-managed workflows and rigorous systems to manage data integrity and security. GIS also delivers contemporary information products and services to stakeholders, taxpayers, and citizens. GIS supports open standards, cloud computing, and extensive worldwide data and high-resolution imagery, providing the essential infrastructure backbone for successful land administration systems.

This book is organized in four parts plus a final section that suggests what readers can do next in using GIS for land administration:

Parcels and property

Part 1 introduces GIS-driven land parcel and property management and the concepts of a digital parcel fabric, which provides a single, authoritative source for land information within a multiuser environment. Real-life stories, or case studies, show how different agencies use GIS to aggregate and standardize parcel data and handle unusual land requirements through visual inspections and edits. These agencies also use GIS for researching changes to a parcel, tracking its history across multiple edits, and sharing parcel maps and applications with other departments, agencies, and the public.

Land management

GIS mapping and analysis provide vital documentation about land information, including land records, tenure, value, and use. These capabilities allow agencies to visualize and share the relationships between land and people. The stories in part 2 show how land management agencies use GIS to create an authoritative geospatial framework for maintaining data integrity, ensuring proper and legally defensible survey records, and tracking land and parcel history. GIS is used to analyze land data to uncover and visualize patterns in land use and its relationship with other issues, such as vulnerable ecosystems, development plans, and natural features of Earth's surface.

Climate and conservation

Part 3 shows how organizations use geospatial tools to anticipate and respond to the impacts of climate change on their communities and operations. With GIS, land and property managers can expose climate risk and the path to sustainable prosperity by providing a transparent, science-based approach. GIS mapping and spatial analysis are critical tools for predicting environmental impacts, planning remediation and conservation, and gaining political and public support for projects related to climate change and conservation.

Equity and social justice

Land is often the cause for misunderstanding and mistreating people. Prejudice, politics, war, catastrophe, and greed can displace people from the land they've occupied for generations or subjugate them to blighted areas and refugee camps. New development can destroy neighborhoods and degrade the quality of life for some while elevating it for others. The stories in part 4 explore how some organizations use GIS to reconnect people with the land of their ancestors and protect the landscapes they love.

HOW TO USE THIS BOOK

THIS BOOK IS DESIGNED TO HELP YOU ADD SPATIAL reasoning to decision processes and operational workflows. It is a guide for taking first steps with GIS and applying locational intelligence to common problems. Using the information from this book can help you create a more collaborative environment within your department and throughout your organization. You can use this book to identify where maps, spatial analysis, and GIS apps can support your work and then, as a next step, learn more about those resources.

The final section of this book, "Next steps," offers readers a basic strategy for applying GIS to land administration.

Learn about additional GIS resources for land administration by visiting the web page for this book.

go.esri.com/vpap-resources

PART 1

PARCELS AND PROPERTY

A PARCEL OF LAND, SOMETIMES REFERRED TO AS A LOT, a plot, or a tract of land, has a legitimate location and surveyed boundaries that a government agency recognizes for the purposes of taxation and development. For community planners, a parcel can represent designated land use, such as residential, commercial, or industrial use, and zoning, which more specifically defines and regulates the types of permitted uses and development requirements or guidelines. For real estate agents and developers, a parcel often represents property and ownership, where the value of the parcel can fluctuate based on market drivers, such as demand and location, and ownership can be transferred or sold. Land administration agencies, government staff, consultants, and real estate professionals must find an efficient way to coordinate their activities and serve their varied interests and legal requirements.

Taxation and ownership are the key components of property records. In the United States, local governments, such as a county government, usually administer property records. However, many departments, agencies, and organizations create, use, and regularly interact with property data, where unstandardized property information and varying degrees of accuracy and precision can lead to misinterpretation, errors, and editing or updating mistakes. The use of GIS can solve parcel data inconsistencies by mapping parcels to a standardized and comprehensive framework, or fabric, for managing, editing, and sharing parcel data in a multiuser environment. The parcel fabric is maintained using a digital or cloud service that allows administrative agencies to share parcel information across devices, enabling a variety of users, including fire, police, and emergency management departments, to do their work based on a single, authoritative source. Additionally, the use of GIS can broaden access to other frequently used land management information and legal records, such as development plans, plats, deeds, and survey records. Land transactions are recorded and stored within the parcel fabric, providing a complete history of changes, and offering users access to the most current status of every parcel.

Perhaps the greatest benefit of applying GIS to parcel and property management is the added ability to analyze and visualize parcel data within the context of real-world significance, including contributing factors, such as soil types, demographics, and economic opportunities and constraints. GIS helps users view and investigate parcel data in 2D and 3D maps displayed as time-series map applications, showing changes in growth, valuation, land use, and development over the span of days, months, and years.

Real-life stories

The first real-life story, or case study, explains how a state agency created a standardized, nearly statewide database of basic land parcel information and address points. The second story illustrates how a city uses GIS to address its unusual setback requirements by creating, editing, and displaying setback requirements visually, ensuring that planning staff and residents apply correct setbacks to each property. In the third story, a city planning department applies GIS to research any parcel, out of hundreds of thousands, and track its history across multiple edits to increase understanding of when and why certain actions took place. The last story in this part shares how a consulting agency used GIS to determine the locations of known historic industrial buildings, as well as previously unrecorded properties, and simplify the process of evaluating properties for historic preservation.

TRANSFORMING PARCEL DATA

Ilyanna Kadich, Texas Water Development Board

WITHIN JUST FOUR YEARS, THE TEXAS NATURAL RESOURCES Information System (TNRIS), a division of the Texas Water Development Board, created a standardized, nearly statewide database of basic land parcel information and address points. At the project's outset, the availability of digital land parcel information from county appraisal districts was restricted for a variety of reasons. TNRIS made this data and address point data readily available as online services through its data hub. The TNRIS mission brings together the Texas GIS community at all levels statewide to identify and prioritize geospatial data needs; establish consistent standards to facilitate interoperability; and serve timely, accurate, and accessible data.

The Texas Geographic Information Landscape report by the state's geographic information officer (GIO) identified land parcels and address points as high-priority datasets, effectively launching TNRIS on its journey to aggregate and standardize all parcel and address point data and make it publicly available. Land parcel and address points are fundamental to government activities such as state lands management, transportation, water resources, public safety, and emergency response.

Data aggregation requires transforming data from multiple sources, including appending, standardizing, and analyzing geospatial data to create various outputs, such as geospatial datasets. Data transformation, the changing of data values and how a dataset is structured (schema), can be intimidating. TNRIS accomplished the transformation of datasets using Esri's Community Data Aggregation tools (an ArcGIS® Solution). The solution includes an ArcGIS Pro Add-in called Data Assistant, which is used for aggregating land parcel and address points and migrating data from a source dataset to a target dataset with different schema. The schema of target datasets for the project was developed by TNRIS with input from more than 30 Texas stakeholders and informed by industry standards and best practices.

The Data Assistant tool saved TNRIS time and money by accelerating the processing of hundreds of datasets. Also, the source-target configuration file in Data Assistant facilitates the annual statewide land parcel and address point dataset update via the TNRIS Strategic Mapping Program, or StratMap, which covers land parcel data for more than 245 appraisal districts and approximately 10 million address points across Texas.

By the beginning of 2020, the TNRIS DataHub provided statewide coverage of land parcels and address points. DataHub datasets are accepted as authoritative sources and are standardized into a common schema. The datasets are shared with the public.

This story originally appeared as "An Easier Way to Transform Parcel Data" by Ilyanna Kadich in the Spring 2020 issue of *ArcUser*. All images courtesy of the Texas Water Development Board unless otherwise noted.

SHAPING SETBACK REQUIREMENTS

Jordan Baltierra, City of Newport Beach, California

MANY MUNICIPALITIES HAVE DETAILED AND ENIGMATIC zoning codes, which can be challenging for homeowners to navigate when they want to renovate or expand their homes. In the city of Newport Beach, California, the zoning code contains more than 300 pages of complex development standards that apply to a tapestry of subdivisions that have popped up since the city's incorporation in 1906. The city's zoning code also includes standards for building setbacks—the distance a structure or part of a structure is set back from the property line. Setbacks can differ from one residential district to another.

In the United States, most cities adopt their setback requirements through zoning codes to regulate building site areas and control what their streetscapes look like. Setbacks are usually classified by a parcel's property lines—the front, sides, and rear. For instance, a standard low-density residential property with a rectangular shape would have a setback of 20 feet in the front, 3–4 feet on the sides, and 10 feet in the back.

In Newport Beach, however, this standard is not uniform. Some areas have alternative setbacks that have been codified into the municipal code. Thus, the city has varying setback standards for parcels based on factors such as the width and shape of a lot, how close it is to the harbor, and its adjacency to an alley. The City of Newport Beach's zoning code also references 59 individual planned communities that contain their own specific setbacks and standards. Additionally, there are several unique properties in the city that don't meet any of these standards. These properties require city staff to review and define site-specific setback determinations and then get them approved by the city's planning commission.

Although Newport Beach's zoning codes can be overwhelming, the city uses GIS to create, edit, and display setback requirements visually. This ensures that planning staff and residents are applying the correct setbacks to each property. The city maintains setback data dynamically via GIS, giving city staff and residents different ways to access it. For example, the city GIS team immediately releases new setback data using a web map viewer application, created by Esri partner Latitude Geographics, so planning staff can perform quality control on the data and residents can view setback information for individual properties.

The City of Newport Beach shares property setback dimensions and data on a web map viewer created by Esri partner Latitude Geographics.

Newport Beach has more than 29,000 parcel features, and approximately 26,000 of them are zoned for residential use. That means that 26,000 parcels require setback information. GIS is used to overcome the challenge of creating and editing the geospatial dataset for setbacks by accurately aligning setback boundaries with property boundaries.

Another challenge is determining which side of the property is the front, side, or rear setback. In general, the front of a house is normally the property line that faces the street, but in Newport Beach, there are some areas where this is not the case, such as when a house butts up against the beach, a cliffside, or the harbor. In addition, many properties have more than one front setback.

Parcels are not static features. They get merged, split, and reshaped over time. Many of the city's GIS layers are dependent on the ebb and flow of this parcel base. For this reason, Newport Beach's GIS team performs continual parcel data maintenance. Additionally, the GIS team created automated spatial queries to find geometric inconsistencies between parcels and relevant setbacks, which enables staff to locate and address each inconsistent record individually.

For Newport Beach, bringing its legacy setback data into a centralized GIS feature class has benefitted planning staff and residents alike. The setback layer bridges complicated zoning policy with comprehensible cartographic visualization.

This story originally appeared as "Setback Mapping in Newport Beach, California, Gets an Update" by Jordan Baltierra in the Winter 2020 issue of *ArcNews*. All images courtesy of the City of Newport Beach, unless otherwise noted.

TRACKING LAND-USE EDITS

Jay Strahan, Department of City Planning, Los Angeles, California

URBAN PLANNING DEPARTMENTS DIRECT GROWTH WITHIN their cities. Through regulations and permit approvals, planners evaluate land use and manage development. Naturally, changes to regulations can cause land-use conflicts, which developers are likely to challenge. That is why planning departments need to have a full record of changes to land use and zoning for parcels, complete with supporting documents that verify that regulations were applied correctly.

The City of Los Angeles Department of City Planning simplified this process of keeping complete records. For more than 10 years, it has used ArcGIS Workflow Manager, Esri's job tracking system, for database editing to ensure data quality.

With Workflow Manager, users can trace edits and link to parcel documentation, enabling the department to maintain crucial metadata about edits in more detail than traditional geodatabase archiving alone. Additionally, because Workflow Manager is an extension for ArcGIS Desktop, staff interact with the data via the normal ArcGIS interface—a program they already know. When they select an area of interest, planners see all the job tracking identifiers associated with that location. From there, they can dive deeper to look at scanned maps, ordinances, history, and why any changes were made.

"ArcGIS Workflow Manager helps us respond to the most important people we serve: our citizens," said James VanGerpen, director of the planning department's Information Technologies Division. "The system allows us to research a parcel and its history across multiple edits and understand when and why certain actions took place. With Workflow Manager, we reliably answer citizens' questions about parcels within an appropriate time frame and reliably inform the decision-making process."

The Los Angeles planning department keeps track of more than 880,000 parcels in 58,000 zoning polygons and 53,000 general plan land-use polygons. Managing this database used to be overwhelming due to the number of changes made to the zoning and land-use datasets throughout the years.

Before implementing the digitized job tracking system, GIS editors' work assignments were based on tiled coverages in ArcGIS Desktop—a total of 1,916 for the city. But supervisors could not decipher or trace edits made to the data. In the planning department's early days, cartographers marked the changes they made to zoning and land use on paper maps, so if someone else took over the assignment, that person would have to follow a paper trail to find clues about ongoing tasks.

Around 2005, department leaders decided to modernize the data management system by implementing a GIS job tracking system. They wanted the technology to make editing more efficient, ensure healthy data, trace previous transactions, and have the agility to evolve and grow. With managers and directors driving the initiative, paper workflows were scrapped in favor of digital ones and staff were reorganized to improve data flows. This leadership-driven flexibility resulted in a successful implementation.

Early in the implementation phase, Esri staff and consultants from Esri partner NorthSouth GIS LLC (NSGIS) worked with the planning department's GIS team to configure Workflow Manager, so it matched the department's editing requirements. The team created automated workflow standards to ensure data continuity, which involved adding custom tools to Workflow Manager using its API.

As technological capabilities advanced, the City of Los Angeles continued to work with Esri to improve editing performance.

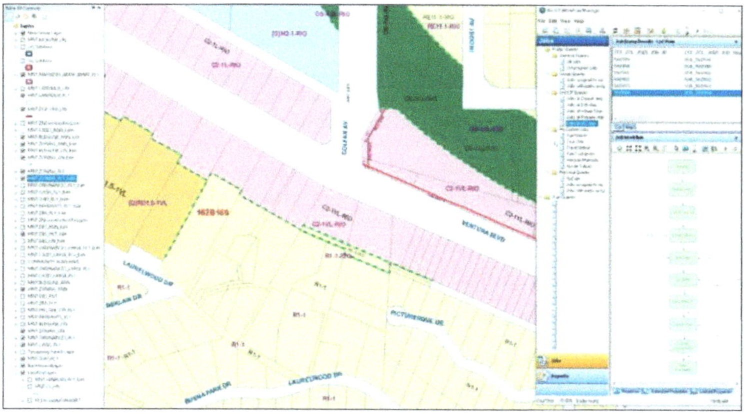

The Los Angeles Department of City Planning uses ArcGIS Workflow Manager to track edits made to data in its three most critical layers: Zoning, General Plan Land Use, and Centerlines.

In 2009, each GIS specialist performed 700–800 edits in Workflow Manager on three tracked layers. As of 2019, they each performed about 2,000 edits per year using automated workflows for editing tasks, version management, geoprocessing, and archiving. The workflows created a version for each job or user, cleaned up old versions, and reconciled and posted changes. What's more, the centralized system streamlined transactional work processes and tracked feature edits. Each edit job took, on average, 5 to 10 minutes to complete.

The planning department maintained 68 datasets that contain information on administrative planning boundaries, city planning case details, and more. Workflow Manager tracks data for the three most critical layers: Zoning, General Plan Land Use, and Centerlines.

The information in these layers is critical to public interest, since it informs how areas of the city can be used and developed. The department makes zoning and property information publicly available via the Zone Information and Map Access System (ZIMAS) website. To ensure the integrity of public-facing spatial data, editors and supervisors from the planning department visually review, correct, and verify data quality. They rely on the repeatable workflows in Workflow Manager to establish data standardization and consistency.

Each year, the planning department tracks an estimated 10,000 data edits. The GIS team creates weekly status reports for the IT directors that show all the information associated with each job. A custom reporting tool aggregates the number of edits made to each dataset, tracks the types of edits (for example, whether information gets inserted, updated, or deleted), and creates a full history of actions by recording each job that is edited within the three tracked layers.

The department also launched Re:CodeLA, a comprehensive revision of the city's existing zoning code system, which has been in use since 1947. The initiative seeks to make it easier for urban areas to evolve and change, and it requires the planning department to rezone every parcel throughout the city. The GIS team anticipates that the new zoning code will triple the number of zones in its database. Workflow Manager is instrumental in assigning tasks to each editor to make the implementation a smooth transition and maintain continuity in the rezoning process.

This story originally appeared as "City of Los Angeles Tracks Land-Use Data Edits, Enabling Smart and Agile Growth" by Jay Strahan, in the Fall 2019 issue of *ArcNews*. All images courtesy of the City of Los Angeles Department of City Planning, unless otherwise noted.

REDISCOVERING HISTORY

Angela Hronek and Sebastian Renfield, Mead & Hunt

ONCE CALLED "THE MACHINE SHOP OF THE WORLD," Milwaukee, Wisconsin, is still home to thousands of industrial properties, including breweries, factories, warehouses, mills, tanneries, and foundries. Many of these buildings and neighborhoods are potential candidates for listing in the National Register of Historic Places, which qualifies them to receive Historic Tax Credits that encourage redevelopment. But the Wisconsin State Historic Preservation Office (SHPO) needed to know where all of Milwaukee's prospects were.

SHPO hired the cultural resources team at Mead & Hunt Inc. to conduct an exhaustive survey of the city's historic industrial buildings. Both parties knew they would need a GIS-integrated approach. So, Mead & Hunt's team, which included architectural historians and GIS specialists, used ArcGIS to review and record large numbers of historic resources in the field and then seamlessly transfer that data to SHPO.

To provide comprehensive survey coverage, the project team needed to identify industrial properties within Milwaukee's city limits that were more than 40 years old. But instead of going straight into the field, Mead & Hunt used GIS data from the City of Milwaukee, along with other existing data sources, to first determine the extent of the survey area and then build field survey maps. The objective was to determine the locations of known historic industrial buildings, as well as previously unrecorded properties.

"GIS analysis of the existing datasets refined and narrowed both the geographic scope and the temporal extent of the data based on project parameters," said Brauna Hartzell, a GIS analyst at Mead & Hunt.

To begin the process, Mead & Hunt obtained the Wisconsin Historic Preservation Database from SHPO, which contains points for previously recorded historic industrial properties throughout Milwaukee. The project team then worked with SHPO to come up with appropriate industrial activity categories—such as *tannery*, *brewery*, and *manufacturing*—and mapped all the properties in the database using ArcGIS Desktop. Mead & Hunt found 546 already-recorded properties that fit into these categories.

Next, Mead & Hunt identified additional industrial buildings that had not yet been documented. Using Milwaukee's GIS-based parcel data, the project team found all the lots in the city that were zoned for industrial use. The plan was to have surveyors review these parcels in the field to determine if there were any other buildings that fit the survey criteria. This review produced more than 1,400 industrial parcels that needed to be added to the survey.

Mead & Hunt overlaid those 1,400 parcels with the 546 previously surveyed features to define the combined survey area, which enabled the project team to pinpoint individual industrial properties and concentrations of industrial buildings that had the potential to be included in the National Register of Historic Places.

Once Mead & Hunt identified all the properties that needed to be surveyed, the company's GIS specialists used ArcGIS Collector to produce tablet-based maps that included every single property. A point layer containing the previously recorded industrial resources served as the foundation for data collection; surveyors would use it to revisit those properties and confirm that they still existed. A second layer based on the 1,400-feature industrial parcel data would then allow surveyors to track their progress by coding each parcel as *visited*, *vacant*, or of *nonhistoric age*.

Along the way, if surveyors encountered any additional industrial buildings on those parcels, they used Collector to add new features to the point layer. The entire survey was conducted offline so surveyors could compare and check the data before uploading it to ArcGIS Online at the end of each day.

Mead & Hunt sent teams out into the field with Collector and took just 10 days to visit all 1,946 properties that had been flagged for the survey.

To discern whether an individual industrial building was a likely candidate for the National Register of Historic Places, the field teams scouted properties that looked exceptionally historic or had

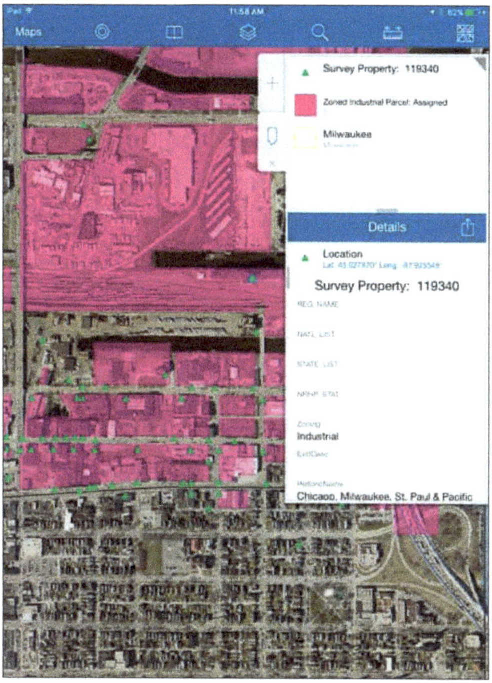

Using Collector, survey teams recorded data on historic industrial properties in the field and then conveyed it directly to the Wisconsin State Historic Preservation Office as digital deliverables, including GIS data and digital photographs.

interesting architectural features. They also searched for places that appeared to be associated with important Milwaukee manufacturers—including former automobile (and current water heater) manufacturer A. O. Smith; clothing and textile producer Eagle Knitting Mills; and the Allen-Bradley Company, which got its start producing electrical controls. The buildings ranged widely in their industrial use, from manufacturing beer, boots, and steam engines to producing candy, caskets, and straw hats.

Many potentially significant properties were less obvious, however. Using ArcGIS Desktop, the project team displayed the survey data on a map and symbolized the buildings based on when they were built, the types of industries they were used for, and other attributes. This helped the team see concentrations of historic buildings and identify trends that revealed the context of industrial development throughout Milwaukee. Having the capability to visualize all this information on a map allowed Mead & Hunt to identify potential historic districts.

Data-gathering using Collector streamlined fieldwork operations, enabling surveyors to track their progress and easily navigate from one industrial parcel to the next. The process also offered several other advantages over traditional pen-and-paper survey methods. For example, developing the survey form in Collector allowed Mead & Hunt to preserve the schema that the Wisconsin Historic Preservation Database uses. This, in turn, enabled SHPO to easily reintegrate the survey records into its own database, eliminating the need to manually enter descriptive data from individual paper forms. Similarly, SHPO obtained property locations without having to digitize each point.

"These new GIS-based survey methods allowed the project team to significantly increase our efficiency and reduce the amount of time needed to prepare for [the] field survey and manage survey data," said Emily Pettis, a project manager at Mead & Hunt. "The client

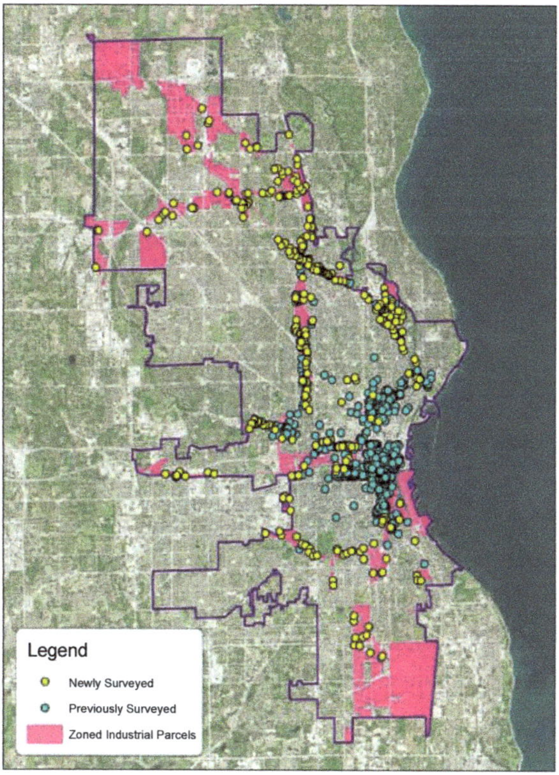

Mead & Hunt mapped data from the City of Milwaukee and the Wisconsin State Historic Preservation Office that identified zoned industrial parcels (pink) and previously surveyed industrial properties (blue). The company also mapped newly surveyed buildings (yellow).

was very impressed with our up-to-date progress maps and digital deliverables, including GIS data."

By the end of the survey, the project team had a spreadsheet that could be seamlessly incorporated into the Wisconsin Historic Preservation Database. The team also had a GIS point layer containing all the surveyed properties, which SHPO uses in the interactive maps embedded in its online database. The results of the survey provide a comprehensive overview of Milwaukee's industrial heritage.

Multiple local developers and nonprofit organizations are already using them to assist in identifying industrial properties that are suitable for redevelopment.

Through the survey, Mead & Hunt distinguished 39 historic buildings and two historic districts—the Layton Park Industrial Historic District and the West St. Paul Avenue Industrial Historic District—as prospective candidates for listing in the National Register of Historic Places. These properties could undergo significant rehabilitation with help from Historic Tax Credits, which would make it easier for local developers to turn the buildings into apartments, offices, and other commercial entities. Since its inception in 1982, the Historic Tax Credit program has reinvested more than $526 million in Milwaukee's historic properties—hopefully with more to come.

Large-scale survey projects, like the one Mead & Hunt conducted for SHPO, greatly simplify the process of evaluating properties for historic preservation. ArcGIS technology makes the fieldwork more efficient and streamlines the digital deliverables, creating more redevelopment opportunities and enabling cities to understand more thoroughly the contexts from which these opportunities emerged.

This story originally appeared as "Milwaukee's Industrial Past Is Rediscovered and Redeveloped Using GIS" by Angela Hronek and Sebastian Renfield in the Winter 2019 issue of *ArcNews*. All images courtesy of Mead & Hunt, unless otherwise noted.

TAKEAWAYS

WHEN IT COMES TO PARCELS AND PROPERTY, LAND administrators, city planners, and real estate professionals have different objectives but share a common interest in the most current and reliable means for accessing and sharing land information. GIS mapping and applications help land administrators track the ever-changing disposition of parcels and property over time, including edits to boundaries, ownership, tax rate, and land-use requirements. Through data aggregation, GIS simplifies navigation through the complex maze of regulations, valuation, and historical records, which otherwise can be difficult to manage across different systems, departments, and agencies.

- **Document changes:** GIS facilitates updates and edits to land records, giving land administrators and technicians the ability to visually review, correct, and verify data quality, while maintaining data standardization and consistency. GIS also connects with ancillary data and assets, such as scanned maps, ordinances, parcel history, and documenting why changes were made.

- **Map unique properties:** Creating a spatial inventory of unique or special case properties allows city, state, and national governments to assess and designate those properties as historical or culturally significant places, such as listing a property in the National Register of Historic Places or qualifying it to receive tax credits.

- **Provide access:** GIS applications provide clear, accessible, and easy-to-use parcel and property information to citizens, businesses, and government departments, which increases transparency, improves understanding of service-level agreements, and supports national-level planning, resource allocation, and disaster response.

- **Analyze property valuation:** GIS enables land administration agencies to analyze and visualize factors that impact property values, including development restrictions, population characteristics, and access to services. With GIS spatial analysis capabilities, land managers and realty companies can refine valuation calculations for more accurate assessments.

PART 2

LAND MANAGEMENT

LAND MANAGEMENT AGENCIES PROVIDE AN AUTHORITAtive framework for describing the recommended use of land and are often the last word on who has ultimate jurisdiction and responsibility. However, not all countries have a unified land management system, or the system they do have is insufficiently implemented to serve as a general guide for land-use decisions. Without a fully realized land management system, land dispute resolution and the definition of proper use are disconnected from authority or a legal process for ensuring implementation, which inevitably thwarts economic and social prosperity and the protection of vulnerable ecosystems.

GIS mapping and analysis provide vital documentation about land information, including land records, tenure, value, and use, and enable agencies to visualize and share the relationships between land and people. With a comprehensive GIS-based land administration system, often referred to as a cadastre system, governments can improve land information management, including property valuation, and provide detailed communication about land and property to government agencies, legal systems, and the public. Additionally, precise mapping of land can define the underlying composition of land, such as soil health, and agricultural viability, such as the suitability of grazing lands. Land development plans and projects inevitably encroach on public lands, farmlands, and other types of landowners. Using GIS, land management agencies can find ways to mitigate the negative impacts of development and ensure fair compensation for landowners.

The core of a modern land administration system is parcel management. A parcel is land with defined boundaries based on survey measurements. Each parcel is associated with additional information, such as a unique identifier, land-use designation, ownership information, assessed value, tax rate, and remote sensing imagery. In GIS, an entire system or network of parcels is called a parcel fabric, which is designed to maintain data integrity, ensure a proper and legally defensible survey record, and track parcel history. With parcel management, GIS is used to create, edit, and securely share land parcel data as map books or online maps; maintain legal descriptions; and publish land data for public and commercial consumption.

Real-life stories

These next real-life stories provide examples of how different land administration agencies use GIS to map and assess land-use patterns using modern technologies, such as the first story about using machine learning to digitize land-use features and characteristics from satellite imagery. In the second example, agencies use GIS to provide better land information services, such as web map services, to local populations, and identify and eliminate errors in land titles. In the third story, a national mapping agency creates and maintains an extensive inventory of public parks and other protected areas for billions of acres of land and marine regions, sharing that information with thousands of agencies and nongovernmental organizations. Finally, a nonprofit organization uses GIS to provide decision-makers with an assessment of the impacts of deforestation and losing topsoil through erosion.

AUTOMATING LAND CLASSIFICATION

Department of Environment and Science, Queensland, Australia

THE STATE OF QUEENSLAND, IN NORTHEASTERN AUSTRALIA, is remarkably geographically diverse. It includes coastal rain forests, widespread eucalypt and acacia woodlands, tropical savannas, ephemeral inland rivers, deserts, and rich agricultural belts. With an area of 1,730,000 square kilometers (668,000 square miles), it is approximately seven times the size of Great Britain.

To map and assess land-use patterns and changes throughout the state, Queensland's Department of Environment and Science (DES) formed the Queensland Land Use Mapping Program (QLUMP) more than 20 years ago.

"Land use has been identified as a foundational spatial dataset that the government considers vital for the progression and development of Queensland," said Andy Clark, senior scientist at the DES Remote Sensing Centre. "The state is large, and it is important that we continue to improve the speed at which we collect this data, as well as its accuracy. In addition, we must keep our procedures in accordance with the standards established by the Australian Land Use and Management Classification system so that it is consistent with data collected throughout the entire country."

Traditionally, the methodology used to keep QLUMP up to date relied on a team of skilled spatial scientists to manually digitize land-use features from satellite imagery. Because of the size of Queensland, this process took a lot of time and resources. "Previously, we have made various attempts to automate QLUMP, all of which ultimately proved to be unsuccessful," said Clark. "Decision tree models were used to infer land-use features from ancillary data; however, this method did not provide an accurate representation of what was on the ground. We tried using the spectral information from satellite imagery to conduct a supervised classification but determined that this procedure could not successfully distinguish between features because, spectrally, they appear very similar. Also, object-based image analyses tended to be just as resource-intensive as manually drawing land-use features."

In recent years, however, machine learning—a subdiscipline of artificial intelligence (AI)—has progressed to the point that using computer vision and deep learning in image analysis and classification is now viable. "With advanced programming tools and computer hardware, the speed and capabilities required to successfully apply machine learning to accurately classify large areas of land looks very promising," said Clark. His team developed a model that can automatically classify different types of land use throughout Queensland.

The Queensland Department of Environment and Science created a computer model that uses machine learning to map banana plantations in the Johnstone River catchment.

Training a model to do machine learning requires a lot of information. Fortunately for DES, it had been using QLUMP for years to collect data. "It was just a matter of refining it so that it could be used in the machine learning process," said Clark. He used ArcGIS Pro and ArcPy™ to generate and refine the training data and applied a range of geoprocessing tools to postprocess the prediction probability from the computer vision model. The Reclassify tool converted the prediction to a binary raster, he explained. The Raster to Polygon tool converted the data to a feature class. The Union tool was used to derive change. And the Eliminate tool merged small features into larger ones.

The QLUMP team independently verified the accuracy of the process by randomly generating thousands of points and assessing the land use at each point. "ArcGIS Pro also generates the error matrix for us, as well as the creation and publication of web maps, apps, and reports to communicate with our stakeholders," Clark added.

In the machine learning process, DES uses a convolutional neural network (CNN) based on a U-net architecture to help the model visually recognize land cover. CNNs are algorithms that mimic the functions of the human brain. By being exposed to large amounts of visual data, the model can learn to distinguish between similarities and dissimilarities in the data. "We borrowed the idea from Olaf Ronneberger, who developed it for biomedical image segmentation, which is a way to identify cells in microscopy images," Clark explained. "Using this architecture, we created an algorithm with 87,153,153 parameters."

The team iteratively fed thousands of satellite imagery patches through the neural network to produce a prediction. The algorithm then self-evaluated and refined the prediction, and the cycle was repeated until it ultimately achieved a 97 percent accuracy rate.

"Python was used to develop the computer vision part of the project," said Clark. "We used NumPy, a library with a large collection of high-level mathematical functions for Python, to handle the multidimensional array and the Geospatial Data Abstraction Library (GDAL) to read the imagery and convert it to the NumPy array—the format required for the neural network."

GDAL can then take an output array and convert it back to an image. "There was a small component of GDAL reading vector data, but ArcGIS Pro was the main tool for processing vectors," said Clark. "We also used Keras, a Python library, for developing and evaluating deep learning models. TensorFlow ran in the back end. It is an artificial intelligence library for data flow and the creation of large-scale neural networks."

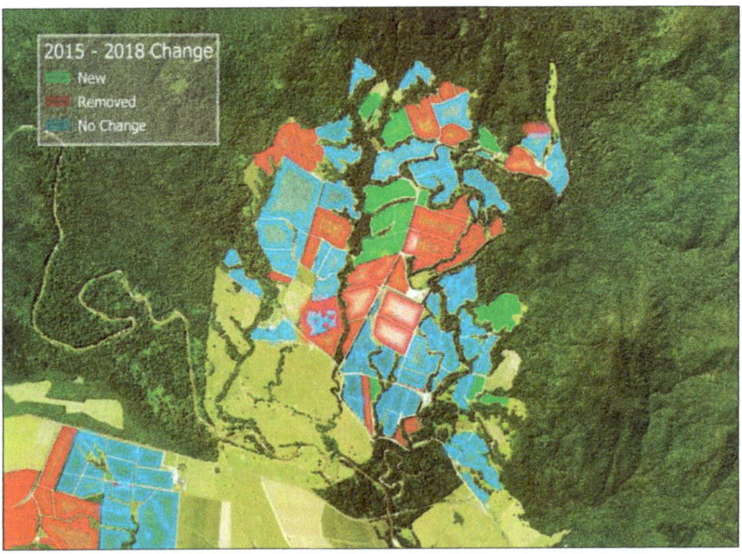

The Queensland Department of Environment and Science used machine learning to model and predict how land use has changed over time. This example shows how a banana plantation in north Queensland transformed from 2015 to 2018.

In machine learning, because of the amount of data that must be processed and refined quickly and repeatedly, processing speed is critical. That's why DES is using eight Tesla V100 graphics processing units (GPUs) that are connected to its high-performance computing infrastructure for deep learning data processing. "The processing speed is amazing," said Clark. "Each GPU essentially provides us with the equivalent of a year's worth of conventional CPU processing in about 2.5 days."

Originally, the model was trained to identify and map banana plantations in the Johnstone River catchment in north Queensland. It was then used to infer banana plantations in the Tully catchment. This allowed the scientists involved in image analysis to focus on the interpretation of the imagery the model produces so they can better inform department decision makers about appropriate biosecurity responses to plant diseases. For example, Panama Tropical Race 4 is a serious disease that can spread rapidly through a banana plantation. In 2015, Queensland's Department of Agriculture and Fisheries (DAF) detected it when examining plant samples. At the time, DES had not implemented its computer vision-based image analysis process. So, determining the potential spread of the pervasive fungus required a team of five scientists per year to manually map and analyze all the banana plantations and other land-use classes in Queensland.

The mapping efforts were subsequently used to help train the CNN model. In 2019, DES received new imagery and updated the banana plantation mapping, which took four days for the computer to complete. Because of its speed and accuracy, the CNN model was trained to map other land-use classes. "Computer vision in fusion with high-performance supercomputing and integrated with ArcGIS represents a paradigm shift that increases our capacity to compile and publish timely land-use information," said Clark. "The methods are sustainable for any image segmentation task and have been applied to mapping wooded vegetation in Queensland, which is quite a different application, as these areas range from dense rain forest to scattered-tree landscapes."

DES intends to expand its methods into most land uses, including other crop types, forestry plantations, and urban land-use classes. "Anything that you can see in the imagery, you can train an algorithm to find. You just need lots of good-quality training data," said Clark. "Based on the methods developed to date, computer vision has the capability to bring efficiencies to large-area mapping and monitoring programs that inform natural resources management and monitoring by governmental and nongovernmental organizations."

This story originally appeared as "Australian State Automated Large-Area Land Classification with Machine Learning" in the Winter 2020 issue of *ArcNews*. All images courtesy of the Department of Environment and Science, Queensland, Australia, unless otherwise noted.

IMPROVING LAND MANAGEMENT

Rwanda Land Management and Use Authority

THE LATEST RELEASE OF THE LAND ADMINISTRATION Information System (LAIS) by the Rwanda Land Management and Use Authority (RLMUA) is transforming the way the country manages, develops, and maps its land.

Begun in 2008, Rwanda's land registration reform has fostered incremental improvements in the Rwanda land administration processes. The first version of LAIS helped support the initial collection and management of large amounts of land administration and legal data for the development of land-use plans at the national and district levels.

In collaboration with Kadaster International, RLMUA used ArcGIS to develop and deploy LAIS version 4.0, which improved the process of titling and certification of land throughout the country.

The land authority is responsible for implementing the National Land Tenure Reform Program, as provided by the National Land Policy and the law governing the land in Rwanda.

During the last two decades, Rwanda has invested considerably in land administration. Most recently, the successful deployment of LAIS version 4.0 in 2019 and the adoption of Esri's Land Administration Modernization Program (LAMP) provided the country with advanced GIS technology and support to meet the evolving challenge of comprehensive land governance.

Esri's LAMP is a grant program that allows resource-constrained countries and their cadastral agencies access to the Esri enterprise platform, supporting the upgrade to LAIS version 4.0 and mapping work.

"RLMUA got multiple benefits from implementing LAMP, such as providing better and [speedier] service delivery to the population; more consistent data; eliminated errors in land titles; and [provided]

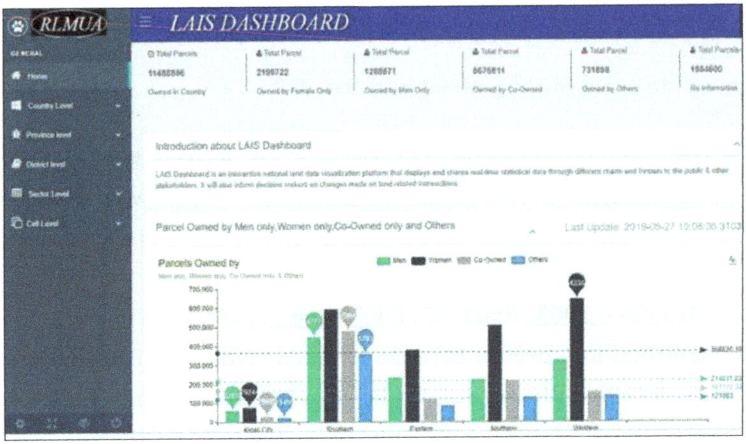

Rwanda's LAMP dashboard integrates interactive national land data visualization for sharing real-time statistics on land use.

a better protection of the land titles [and] web map services, among others," said Espérance Mukamana, director general and chief registrar of land titles, RLMUA.

The new system overcame technical challenges encountered in the previous LAIS versions by combining previously separated textual and GIS components into one system and better aligning with the current IT system's security procedures. The new system also ensures alignment with the ISO 19152:2012 Land Administration Domain Model (LADM), which supports standardization and integration of land data.

LAIS version 4.0 provided users with web access to up-to-date spatial data through mapping services, which can be shared seamlessly. It also integrated interactive national land data visualizations that share real-time statistics on land use through charts available on a central dashboard.

Additionally, the new system enhanced transparency and provided users with the spatial tools and data they needed to perform

Rwanda Land Management and Use Authority uses advanced geospatial technology to operationalize land administration and real-time monitoring of land rights and use for potential investments in the country.

land-use analysis for their own land and, collectively, to better plan for the sustainable development of the country.

RLMUA has gained remarkable experience and results by using GIS for land management services, according to Mukamana, who says the technology has "led to a better understanding and implementation of the land management system and helped RLMUA achieve its targets efficiently."

LAIS version 4.0 places Rwanda among the few countries on the African continent with the level of advanced geospatial technology that can operationalize land administration and use capabilities while also enabling real-time monitoring of land rights and use for potential investments in the country. The entire system has proved to be a vital tool that informs decision-makers on any changes made to land-related transactions.

"This project represents a lighthouse for Africa," said Sohail Elabd, Esri Global Business Development general manager, Middle

East and Africa region. "Many governments are looking at ways to increase revenue through registration of land. Esri technology, provisioned through the LAMP, provides a set of out-of-the-box tools to do so. We are excited to be working with RLMUA to make this happen."

This story originally appeared as "Rwanda Improves Land Management Processes with GIS Technology" in the Spring 2021 issue of *ArcNews*. All images courtesy of the Rwanda Land Management and Use Authority unless otherwise noted.

INVENTORYING PROTECTED AREAS

Lisa Johnson and Mason Croft, Boise State University, and Larry Orman, GreenInfo Network

THE PROTECTED AREAS DATABASE OF THE UNITED STATES (PAD-US) is the official GIS inventory of public parks and other protected areas in all US states and territories. It contains 3 billion land and marine acres managed by more than 15,000 agencies and nongovernmental organizations (NGOs) in more than 150,000 separate parks and protected areas.

Developed and managed by the US Geological Survey's (USGS) Core Science Analytics, Synthesis, and Libraries, the most recent data has data for federal and state agencies. PAD-US also has nonsensitive easement data from the National Conservation Easement Database

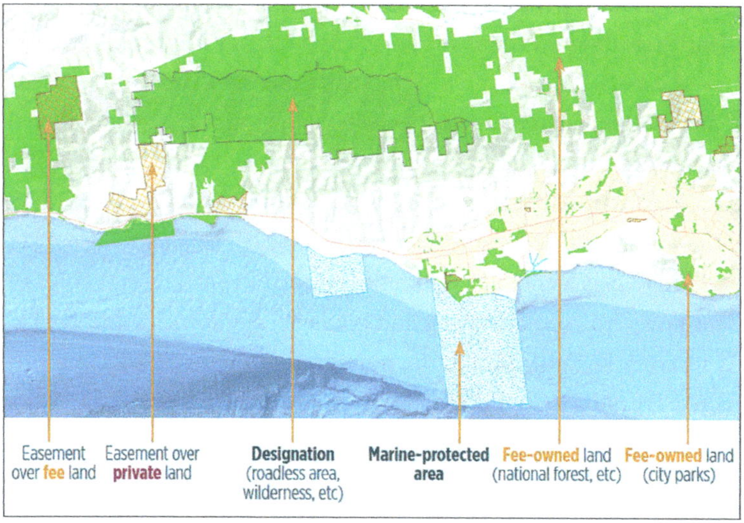

The Protected Areas Database of the United States categorizes fee and private lands, designation areas, easements, and marine protected areas.

and information on marine protected areas from the National Oceanic and Atmospheric Administration (NOAA).

The data in PAD-US can be used to meet a wide range of needs. For example, app developers who work on park finder apps, such as Recreation.gov, or ParkRx apps, which allow doctors to prescribe park use to patients with chronic diseases, can use PAD-US to build their web apps. The National Wildfire Coordinating Group can use the database to effectively manage fire response. Communities and nonprofits can use it to define conservation priorities for habitats and species. And the database offers a framework for tracking federal and state Land and Water Conservation Fund investments.

Regional and county planners can also use PAD-US to better assess park needs in their areas. Additionally, local and state governments can use the maps and data to calculate parks' economic benefits for tourism-based economic development ideas and better guide visitors to local parkland. And land planners can employ PAD-US to identify the best locations for energy and transmission facilities.

The PAD-US database is particularly useful to anyone who needs regional data, including counties; intrastate, multistate, and national organizations; research bodies; and planning and policy groups. The data is also available as an ArcGIS web map, making it easy for Esri users to find and apply it to their work.

PAD-US is a complex dataset developed by integrating scores of data flows from land-owning and management agencies and NGOs. At its core, PAD-US defines the boundaries of public parks and protected areas owned and managed by governments and nonprofits, including terrestrial and marine areas.

To get federal, state, and NGO data into PAD-US, USGS works with federal agencies and data stewards—such as state governments, universities, and nonprofits—to aggregate data for state and local parks and protected areas. With more than 30,000 state, regional,

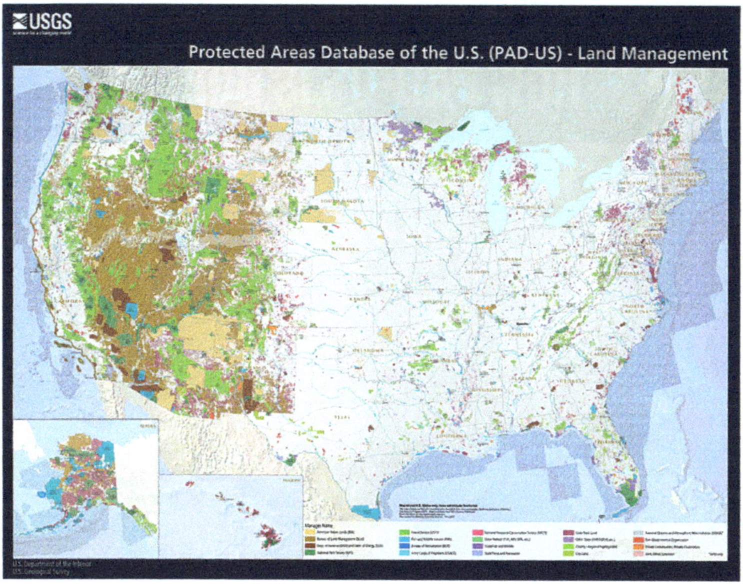

PAD-US contains 3 billion land and water acres managed by more than 15,000 agencies and NGOs in more than 150,000 separate parks and protected areas in the United States and its territories.

local, and NGO land managers, state aggregators create a distributed yet viable system for gathering the data.

The PAD-US development team manages all the data in ArcGIS. Geodatabase domains help PAD-US data managers standardize attributes. Having multiple feature class representations supports data development and managing overlapping designations.

PAD-US web mapping services, which provide users with ready-made interpretations of the data, are available through USGS. The USGS PAD-US Viewer and the web map are examples of how these map services can be used to aid interpretation, with the data divided into Manager Type, Manager Name, Protection Status, and Public Access.

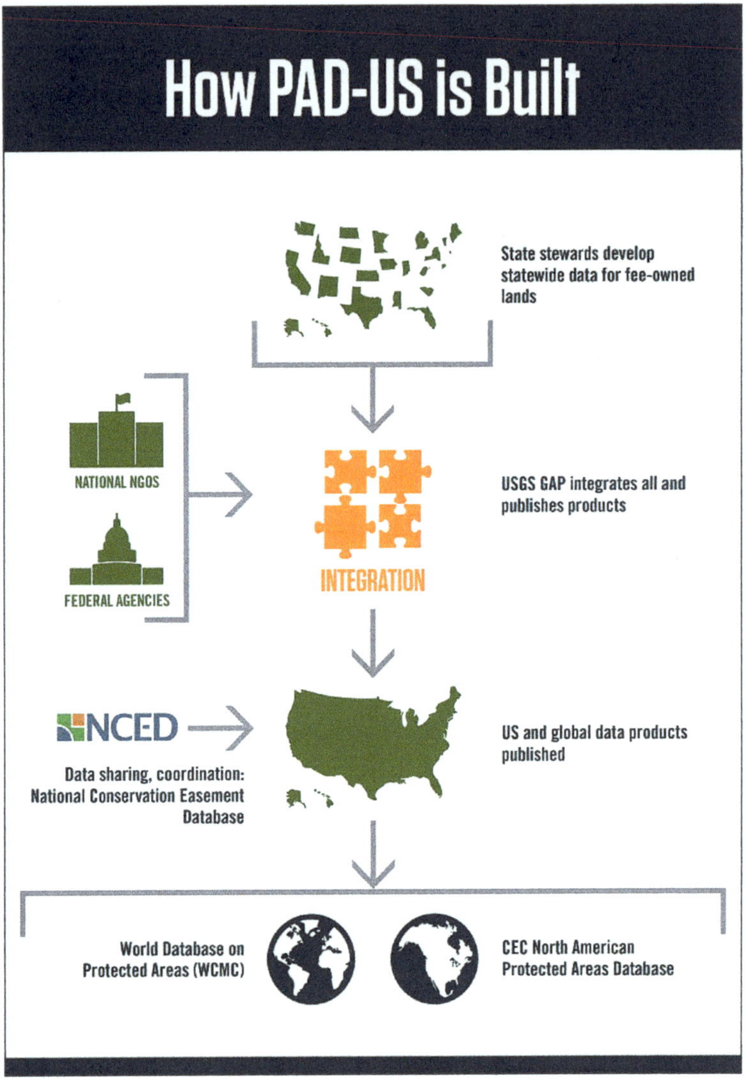

PAD-US is a complex dataset developed by integrating scores of data flows from local, state, and federal land-owning and management agencies and NGOs.

PAD-US is first and foremost a product of collaboration and coordination, particularly at the federal, state, and NGO levels. USGS cochairs (along with the US Census Bureau and the Bureau of Ocean Energy Management) the Federal Lands Working Group, an innovative partnership of 15 federal agencies that define data translation protocols for sharing all federal-level land management data directly into PAD-US.

The state-level data stewards—which include one or two governmental agencies, universities, and nonprofits per state, depending on each state's situation—manage state and local data for PAD-US updates. Stewards range from departments of natural resources and natural heritage programs to universities like Colorado State University and nonprofits such as the GreenInfo Network.

Nongovernmental organizations also play key roles in PAD-US. The Nature Conservancy, for example, has helped develop inventories in the eastern United States. The National Conservation Easement Database, which is led by the Trust for Public Land and Ducks Unlimited, secures easement data from land trusts and agencies. And the National Recreation and Parks Association promotes PAD-US-compatible data standards and tools for its thousands of agency members.

Work has also moved forward on many other technical improvements to PAD-US, resolving boundary gaps and overlaps, filling in data for some attributes, and improving the status ranks for biodiversity conservation.

This story originally appeared as "Building a Complete GIS Database of Protected Areas" by Lisa Johnson in the Spring 2017 issue of *ArcNews*. All images courtesy of Boise State University, United States Geological Survey, and GreenInfo Network, unless otherwise noted.

INVESTING IN SUSTAINABILITY

Anne Cassidy, Winrock International

PRIVATE-SECTOR INTERESTS IN CASH CROPS, TIMBER, AND textile mills compete with Cambodia's need to feed its people and protect its natural resources. Government decision makers must try to balance private investments with sustainable development goals. It has been difficult to compare the potential benefits and costs of development options because decision makers lacked information.

But several years ago, the nonprofit organization Winrock International stepped in with a digital tool it designed using Esri software. The Watershed Ecosystem Services Tool (WESTool) was carefully conceived with detailed stakeholder feedback and consultation to make a real difference in the health of Cambodia's fragile ecosystems.

The WESTool allows users to explore how ecosystem services, land uses, and socioeconomic factors interact across Cambodia's landscapes. By combining advanced science with intuitive maps and tools, the WESTool offers valuable information at the local, regional, and national scale to help decision makers and land managers understand and balance the value of remaining forests with development goals.

Winrock, with more than 100 agriculture, environment, and social development projects in more than 40 countries, has used Esri software for more than 20 years.

WESTool grew out of Winrock's Supporting Forests and Biodiversity project, funded by the US Agency for International Development (USAID). Winrock scientists worked with Cambodians to identify their major development challenges and goals. Winrock then used Esri software to design a tool that could anticipate changes in water availability, sediment and nutrient loss, water pollution, and greenhouse gas emissions to determine the best policies and practices

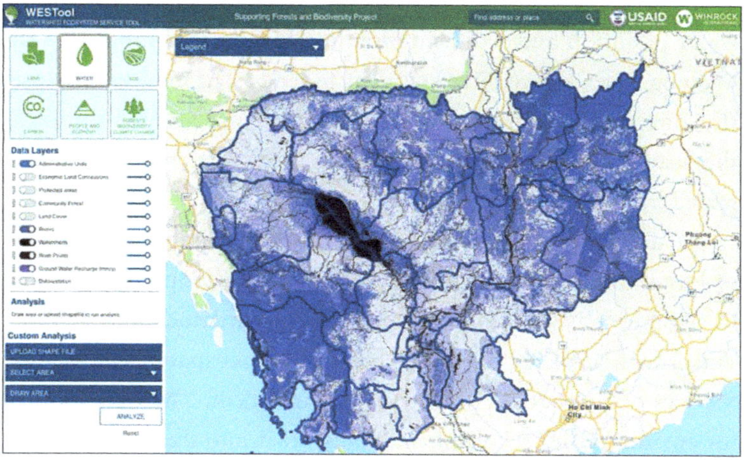

Winrock International's WESTool interface allows users to intuitively select and explore data layers and produce analyses of the impact of development scenarios on key ecosystem services.

for agriculture and land management. By bringing together many different types of spatial information, modeling effects, and presenting this information in a nontechnical format, Winrock scientists created a web-based tool that uncovers the impact of widespread land-use change and suggests strategies to stop deforestation.

"The application is unique because it provides decision makers with a tool to assess the impacts of forest loss on ecosystem services," said Mike Netzer, a GIS and remote sensing analyst with Winrock. "Many decision makers understand there are environmental risks, but they have never had a tool or any way to quantify them. By combining technological advances with a deep understanding of the challenges facing national and local governments, WESTool helps Cambodians make informed decisions to balance economics and the environment." Netzer helped introduced the tool to Cambodians, from government ministers to students.

Cambodia's Ministry of Environment uses the tool to monitor

water pollution and sedimentation in the watersheds that feed the Mekong River, where there are many fisheries. The ministry also uses the tool to predict greenhouse gas emissions from deforestation in the beautiful and protected Prey Lang landscape, where logging is forbidden but still happens.

"The WESTool is a great resource for the Cambodian Ministry of Environment and the Ministry of Natural Resources and Rural Development as we create our state of the environment report," said Vina Touch, director of the Department of Geospatial Information Services of Cambodia's Ministry of Environment. "We plan to use it for environmental reporting and integrate it into the geospatial portal currently under development."

Local communes use the WESTool to create land-use plans and evaluate how dams, roads, and other land-use decisions will affect

Winrock International's WESTool lets users explore hot spots in Cambodia by investigating erosion prevention, aquifer recharge, avoided carbon emissions, and other metrics.

the landscape. "Just by clicking on the tools in the WESTool home page, we can see potential impacts regarding land use, deforestation, drought, or flood modeling," said Ley Phanna, chief of the Planning and Supporting Communes office in Stung Treng Province, Cambodia.

It's difficult to understand the impact of land management decisions without proper, unbiased, science-based information, said Lara Murray, a low-emission development expert with Winrock's Ecosystems Services team. That's what the WESTool provides.

For example, Cambodia's Tonle Sap Lake, one of the world's largest freshwater fisheries, is surrounded by extensive agriculture but also significant remaining forests. WESTool shows that since 2000, more than 19 percent of the forest in that watershed has been lost. This loss has caused an increase in sediment and nitrogen loads in the lake. WESTool users can also look at how future changes will affect the lake.

Just a few years ago, an analysis of this type would have only been done for select locations and select purposes. The results would have only been accessible to scientists and a few government officials. Because it would have required an expensive analysis and likely been localized to a specific area and issue, the analysis would not have considered the overall landscape.

"The point of WESTool, originally and always, was to provide an objective platform for Cambodians to evaluate what is happening to their landscape and the impacts of it," said Dr. Sarah M. Walker, director of Winrock's Ecosystems Services team. "We're at a point in history where we have all this amazing data and all this amazing computer power, and that has finally started to reach other countries below the national level, at provincial and local levels."

Walker said that what technologies like WESTool offer developing nations is nothing short of revolutionary. "Often in the past, things were done because some group was advocating for them to be done, not because it was the most pressing problem. WESTool democratizes the data, placing the power of information into the hands of all the people of Cambodia."

As countries like Cambodia integrate UN Sustainable Development Goals (SDGs) into their development planning, WESTool can also act as a platform to help countries plan, implement, and monitor key indicators. The ability to assess SDG baseline conditions and monitor them over time can improve the potential for multilateral SDG financing, which can catalyze sustainable development.

"Developing countries have to prioritize. They can't just say, 'We're stopping all deforestation.' They need to know where the effects of deforestation and degradation are the worst," said Winrock President and CEO Rodney Ferguson. The Cambodian government can use the WESTool to determine that deforestation has altered the water cycle in the southwestern part of the country and that they are not only losing more topsoil through erosion but also seeing less water infiltrate into the aquifer because of decreased tree cover. These changes show decision makers that forest loss has livelihood and economic impacts. "Then they know that whatever they do, they cannot ignore deforestation and degradation in the southwestern part of Cambodia," Ferguson said.

WESTool has been used to evaluate ecosystem services for 80 percent of Cambodia's land area, home to 8.7 million people. And Winrock continues to improve the tool through its USAID-funded Sustainable Water Partnership, working with Cambodia's Ministry of Environment to promote the tool in environmental impact assessments and commune development plans and to assess water security issues in the Stung Chinit watershed north of Phnom Penh. As users experience including WESTool in their planning processes, Winrock has incorporated additional ecosystem services, problem domains, and economic sectors. WESTool is scalable to other countries and terrains.

In 2016, the Cambodian Ministry of Environment honored Winrock with the Royal Order of Sahametrei—the highest honor the country bestows on foreigners—in recognition for outstanding service to the king and people of Cambodia. Winrock was lauded for conserving Cambodia's biodiversity and forest resources, which included the country's use of WESTool.

What's most exciting about WESTool, say the people who created it, is how it puts huge amounts of usable data in the hands of ordinary people. "It brings people science in a way they otherwise might not be able to access," Murray said.

This story originally appeared as "Helping Cambodia Find Its Way" by Anne Cassidy in the Winter 2019 issue of *ArcUser*. All images courtesy of Winrock International unless otherwise noted.

TAKEAWAYS

LAND MANAGEMENT REQUIRES AN AUTHORITATIVE DIGITAL framework for designating proper land use and responsibility. Managing authorities use GIS to connect land-use decisions with legal documentation to resolve land disputes, support economic health, address social prosperity, and protect vulnerable ecosystems. GIS mapping and analysis provides vital information about land, including land records, tenure, value, and permitted use, and enables agencies to visualize and share the relationships between land and people.

- **Create a parcel fabric:** With GIS, land administration agencies can establish an integrated network of parcel data, a single system for maintaining data integrity, ensuring proper and legally defensible survey records, and tracking parcel history. A parcel fabric is used to define ownership information, assessed value, and tax rate and to connect with remote sensing imagery.

- **Assess land-use patterns:** One of the most powerful capabilities of GIS is spatial analysis. Land administration agencies can analyze land use across multiple variables, such as ownership and permitted use, to uncover and visualize patterns, or against other GIS data, such as vulnerable ecosystems, development plans, and natural features of the earth's surface.

- **Identify and eliminate errors:** Finding land record errors can be difficult when those records are maintained in separate systems. GIS allows land administration agencies to connect systems and consolidate complex information so that the locations of errors, outliers, and gaps can be visualized and addressed comprehensively, reducing conflicts and misunderstandings.

- **Share authoritative land information:** GIS maps and applications are designed to be shared with participating agencies, partners, contractors, and the public. Maps provide an intuitive visual language that land administration agencies can use to improve communication, educate leadership and customers, and convey the most current and accurate information.

PART 3

CLIMATE AND CONSERVATION

PROTECTING AND CONSERVING VALUABLE LAND IS often thought of as an issue that concerns environmentalists, historians, and real estate professionals. However, the value of land is measured along a shifting scale of criteria and concerns that involve many types of people and organizations, which weigh cost in terms of loss, opportunity, and the ever-changing world. For example, sea level rise because of climate change threatens coastal communities and vital ecosystems worldwide. Declining habitats and the fracturing impacts of incompatible land use disrupt wildlife migration patterns and contribute to species extinction. In some cases, the qualities that make land desirable also lead to destructive development of those lands.

Today, conservation technology is disconnected, specialized, and often insecure. Integrated technology systems, like GIS, are desperately needed to manage natural areas and protect them from unprecedented threats such as habitat loss, poaching, and climate change.

Conservationists can use GIS to track wildlife conflicts and engage stakeholders in and around protected areas, using a data-driven approach to manage conflicts between humans and wildlife. GIS-based operations help organizations track poaching, monitor protection plans, and visualize patterns of illegal activity. Spatially augmented conservation easement programs are used to monitor, evaluate, and protect established natural resources such as forests, riparian areas, and wildlife.

Geospatial tools for climate change help leaders anticipate and remedy the impacts of rapid climate change on their communities and operations. Spatial data and GIS can show patterns of climate risk and the path to sustainable prosperity by providing a transparent, science-based approach to this challenge.

Real-life stories

These real-life examples illustrate how people and organizations use GIS to adjust their approach to land management and better understand the multitude of interconnected variables that describe the most beneficial purpose of a landscape. The first story explains how GIS allows organizations to explore and build on new ideas for wetlands and ecosystem restoration as sea levels rise, even within urban environments. The second story details how GIS users can find ways to balance and mitigate conflicting land uses, such as the demand for housing versus the protection of unique and vulnerable ecosystems for flower habitat in Peru. The third story explores how leaders, scientists, activists, and citizens work together using GIS to find the subtle pathways that bond fragile landscapes to each other. The final story explains how the technology makes it possible to run virtual conservation scenarios to identify land best suited for preservation and elevate the level of public discourse about sustainability and preservation.

ADDRESSING SEA LEVEL RISE

Dawn J. Wright, Esri

FLOODING ALONG ONE OF AMERICA'S GREAT estuaries—the San Francisco Bay—prompted local groups to act against further climate risk. The San Francisco Bay Trail provides more than 350 miles of walking and cycling path through the 47 cities and nine counties that ring the San Francisco Bay. The trail also serves as a popular viewing platform for tide cycles and abundant wildlife. Because it sits at the wetland-urban interface, the trail is extremely vulnerable to sea level rise. "In the San Francisco Bay, we have a long history of wetlands and ecosystem restoration for habitat conservation," said Julie Beagle, former senior scientist in the Resilient Landscapes Program at the San Francisco Estuary Institute (SFEI).

To understand the Bay Trail's vulnerability and determine ways to mitigate flood damage, planners with the East Bay Regional Park District use SFEI's *San Francisco Bay Shoreline Adaptation Atlas*, created by Beagle and her colleagues. The atlas serves as a science-based framework for developing adaptation strategies specific to the area's diverse shoreline and that take advantage of natural processes.

"We started to get calls from regulatory agencies dealing with proposals for horizontal levees and beaches—all kinds of new living shoreline ideas," Beagle said. "The regulatory environment, which protects the bay, has historically prohibited the use of sediment to fill in the bay, and there was no science or regulatory guidance on how to soften the shoreline to deal with sea level rise."

The standard method has been to harden the shoreline using riprap and concrete seawalls. Beagle and her team at SFEI set out to share methods and introduce green infrastructure into the region's shoreline management practices. For much of its 40-year history,

SFEI has used GIS to compile information about the estuary's chemical, physical, and biological health. GIS also guides SFEI's mitigation strategies and actions.

Wetland loss continues to be a priority for the San Francisco Bay because an estimated 80 percent of marshes and mudflats no longer exist, according to the California Natural Resources Agency. This loss eliminated wildlife habitat and storm-buffering capacity, which will be vital as storm intensity increases and seas rise. In 2016, voters in the nine Bay Area counties approved a bond measure to provide $500 million over 20 years for wetland restoration. In some areas of the bay, restoration will suffice, but others will require hard infrastructure such as armoring the shoreline and raising levees.

"There are a lot of big questions, and we used GIS to look at the historical knowledge about the bay to determine what makes sense geomorphically," Beagle said. "After Hurricane Sandy, there were a lot of images of marshes being built at the bottom of Manhattan when we know there's deep water there with high wave energy—that's not going to work, and it won't last. We spent five years on the atlas to show the best types of projects for the physical conditions in each place."

Beagle also hopes that use of the Adaptation Atlas will help instill a regional approach spanning all shoreline jurisdictions including cities, counties, wastewater treatment plants, highways, and private landowners. "It's about collaborative planning and collaborative science for resilience," Beagle said. "If each entity is just going to plan for its own piece of shoreline, it's not going to be effective, it won't be resilient, and it doesn't have any ecological benefit. The poorest communities get left behind, and the richest communities pay to build themselves out of sea level rise. That's what we see happening right now."

In addition to the equity issues Beagle raises, nature does not operate according to jurisdictional boundaries or the built environment. "The idea is to capture the processes of watersheds, creeks, and the shoreline to plan holistically," Beagle said. "Another way to think about it is nature's jurisdictions."

SFEI collects many layers of information about the bay, estuary, and watersheds, and it acts as a state spatial data center. All bay restoration projects are entered into SFEI's EcoAtlas to be tracked over time by regulatory agencies monitoring compliance. SFEI purchases aerial and lidar imagery to monitor change and accurately measure elevations. For more thorough or immediate imagery analysis, the institute deploys drones to key areas of interest.

"With the drone, we're able to survey right after a storm or return at the right time of year to see what's changing," said Pete Kauhanen, GIS manager and drone expert at SFEI.

The flexibility of the drone platform allows SFEI to monitor sediment volumes in stream channels and tidal marshes, monitor restoration sites, and capture harmful algal blooms. SFEI also uses drone imagery and a sophisticated machine learning algorithm to count and monitor trash. Many of the drone missions fulfill multiple objectives. "Flying over the Corte Madera Marsh allows us to see patterns of erosion or progradation of the marsh edge, and at the same time we can see the state of trash in the marsh," said Tony Hale, program director of the Environmental Informatics Program at SFEI. "We're able to look at both the built environment and the natural environment and will continue to look back at the record to what has changed."

SFEI's drone team also assisted with the institute's Resilient Landscapes Program in quantifying the impact of winter storms. High-resolution drone imagery and tools that automatically calculate volume provided an accurate measure of sediment and its movement. The Resilient Landscapes team analyzed the imagery to identify erosion and the surge channels that have eaten away the marsh. The work also captured an area with a degrading levee that caused the beneficial expansion of a mudflat.

By using imagery to observe and analyze natural processes, SFEI teams can guide strategy and measure the performance of different management practices. "Imagery also helps us gauge the amount of sediment that we need to supply the marshes so they can keep up with climate change," Hale said.

The Bay Area is home to a broad diversity of people, topography, tidal patterns, and microclimates. Strategies that work in one city for one stretch of shoreline may not work in a neighboring city. The *Adaptation Atlas* cataloged and described the 30 distinct operational land units of San Francisco Bay's 400-mile shoreline. "We look at what makes sense for the land use that's existing and for what could be there in the future," Beagle said. "The *Adaptation Atlas* gives us a framework within which we can work using operational landscape units that are based on our understanding of the underlying geology and geomorphology."

Modeling the processes at play helps SFEI tackle the bigger question of how to improve climate resilience while maintaining functional ecological and hydrologic systems. "By using big spatial metrics, we work to understand how high and wide certain features need to be to knock down waves in each location," Beagle said. That knowledge is then modeled against sea level rise projections to devise adaptation strategies.

From the 1850s to the 1960s, the San Francisco Bay was often filled in to create land for housing and infrastructure. A lot of homes sit on land that's built from artificial fill, which poses its own problems in this earthquake-prone area. Wreckage from the catastrophic 1906 earthquake was simply dumped into the bay and surrounding marshland. Some homes even sit on former garbage dumps that were once open landfills surrounding the bay. "They used to talk about how bad the bay smelled," Beagle said. "Then with the environmental movement and changes in laws, you could no longer fill the bay, and we began cleaning it up."

While nobody can yet claim that the bay is clean, years of remediation have improved water quality and restored vital habitat. "Now, it's less about fixing the problems of the past and more about how we prepare ourselves for the future," Beagle said. "We don't want to wind up with seawalls everywhere. It's about how to have a shoreline that's still ecologically functioning, accessible to people, and equitable."

This story originally appeared as "How an Atlas of San Francisco Bay Is Helping Deal with Sea Level Rise" by Dawn J. Wright in the *Esri Blog*, February 26, 2021.

DEFENDING FRAGILE ECOSYSTEMS

David Gadsden, Esri

HIGH IN THE FOGGY PERUVIAN HILLS, FOR JUST A FEW weeks in June, bright-yellow flowers resembling tiny phonograph horns light up the hillsides. The Flower of Lima, with its naturally short, seasonal life and the fog oasis it calls home, is increasingly under threat.

Illegal land-grabbers seized on the demand for affordable housing as more Peruvians are pushed further from Lima's established settlements. These land traffickers having staked and sold fraudulent claims to the soil that the Amancaes flower (pronounced: ah-mawn-KIE-us) calls home. The housing crisis has led people to build structures on land that not only needs protecting but can be vulnerable to earthquakes and poses health concerns because of the pervasive humidity. The unmanaged development, including unsanctioned quarries where miners extract construction materials, has left behind damaged land unfit for plants.

The unique flora and fauna that make up the fragile ecosystems of this region's *lomas*, or fog oases, have long been at risk of disappearing. A conservation effort in late 2016 led to the formation of the Lomas EbA project, an initiative directed by the National Service for Natural Protected Areas (SERNANP) of Peru and implemented by the United Nations Development Programme (UNDP) with support from the Global Environmental Fund (GEF). The project aims to preserve 19 of the estimated 100 lomas that only occur along the desert coasts of Peru and northern Chile. "We can't lose the only place where this amazing flower grows," said Adriana Kato, a communications specialist with the UNDP group working to preserve the lomas.

The disappearing fog oases in this high-elevation desert are at the crux of a demand for housing at a time when about one-third of Peru's population, nearly 11 million people, live in Lima. While poverty levels in Peru dropped from nearly 59 percent to about 22 percent between 2004 and 2015, much of the population remained in danger of returning to poverty. With a lack of housing strategy in Lima, unorganized development has encroached into the fog oases. "It's not just an environmental problem," Kato said. As a UNDP report noted, the effort addresses an "unprecedented complex combination of problems."

The endangered hills faced a unique challenge: few people have witnessed the land in its full glory because it's covered in a blanket of mist from May to October. By the time the fog lifts, the land turns back to dry desert. Kato, who grew up in Lima, was among those who didn't know about the nearby beauty until she joined the UNDP. "Lima is surrounded by these fog oases, these green hills, but not many people know about them," Kato said. "We have all this, but we don't see it."

The UNDP's Maria Miyasiro, a GIS and remote sensing specialist, raised awareness by building a detailed geoportal to share local apps and maps, including one named GeoLomas, to allow anyone to explore the areas that had gone unseen. Local environmental leaders, often self-taught experts in botany and laws relating to the lands, gathered data for GeoLomas using ArcGIS Survey123 app on phones and tablets to input what they observe as part of a pilot project. "I decided to build a web app so our stakeholders and our environmental leaders could do their own analyses and make their own maps," Miyasiro said. "We started with just one application, the GeoLomas web map viewer, and then we elaborated with more applications created together with the stakeholders, organizations, and municipality workers, and using ArcGIS StoryMaps[SM] to publicize their work."

Volunteers added layers to the GeoLomas map, including archaeological and cultural features, potential tourism locations, and land rights to know when an area was being illegally encroached on. The group worked closely with park rangers to record flora and fauna species they encountered. The online map was described in a recent midterm review of the project as the "biggest advance" toward tracking conservation actions and improving transparency.

The UNDP team also tackled several of the United Nations 17 Sustainable Development Goals: improving the environment for those living on the edge of Lima where there has been less access to green space, creating a sustainable and resilient city, conserving the land and biodiversity of hill ecosystems, and reducing poverty by creating jobs in ecotourism and agriculture. "It's really necessary that people understand that the lomas are not the place to put their houses," Kato said. "It's dangerous because of earthquake risk and for their health because of high humidity. It's better to conserve them, and make a really nice and safe path, so people will visit them and use it for ecotourism, encouraging community members who may have otherwise lived in the oases to instead preserve the lands and launch enterprises for paying visitors."

In an audit of the lomas project's progress, a report noted that despite the complexities faced, the will to participate was positive. In Lima, about 10 different local organizations care for their nearby fog oases, in some cases starting nurseries to nurture new plantings and using a fog-catching mechanism to collect water for irrigation.

Before the creation of the online map, many people didn't know the physical boundaries of the land they were protecting. Now, on a smartphone using the ArcGIS Explorer app or directly in the browser using the GeoLomas web application, they can see whether a house has been illegally erected inside the border or if a new road and plots for sale belong to land-grabbers, requiring them to call the police.

During the COVID-19 pandemic, public outreach and training for contribution to the GeoLomas map project shifted even further online, and leaders increasingly relied on technology. For example, drone imagery and video of the land became a primary mapping feature, helping to raise awareness with immersive video that allows anyone to fly over the scenic hills.

In a little more than four years, the work of organizers and the volunteer force of residents helped map and catalog information about this area. Details gathered provided the evidence needed for formal preservation of at least five of the oases, which are now part of a newly designated regional conservation area called Sistema de Lomas de Lima.

The UNDP team devised a strategy for handing control of conservation efforts back to Lima's government, including bringing housing, cultural affairs, and law enforcement officials to the table. The team encourages private conservation developments and works with the Ministry of Culture to protect other fog oases that include archaeological features at risk of being lost. The team also partnered with the Geological Institute of Peru to preserve unique geologic formations. By collecting key data and mapping it with GIS, advocates communicated the location of each fog oasis, identified local flowers and animal species, identified the sites of nearby archaeological locations, and offered details about pedestrian access. "It's a really, really valuable tool for us," Kato said, relating what GIS revealed—or as she and Miyasiro wrote in the interactive map they authored, "You don't take care of what you don't love, and you don't love what you don't know."

This story originally appeared as "Maps Cut through the Fog in Peru to Help Preserve Unique Ecosystems" by David Gadsden in the *Esri Blog*, May 18, 2021.

CONSERVING CLIMATE-RESILIENT LAND

Eric Aldrich, The Nature Conservancy

THE RESILIENT AND CONNECTED NETWORK, A GIS MAPPING tool developed by more than 270 scientists led by The Nature Conservancy (TNC), gives conservationists a way to save biodiversity for the future. It is available for land trusts, government agencies, and scientists for conservation planning throughout the contiguous 48 US states.

GIS has long been one of conservationists' main tools to identify and protect land, water, plants, fish, wildlife, and habitat across the globe. Using GIS to identify places with high levels of biodiversity has often played a leading role in those efforts.

Conservationists more recently have faced the great challenge of climate change, which is altering biodiversity and habitats. As warmer temperatures, increased flooding, and other climate impacts alter and destroy habitat, species are being forced to find new places to live. Diverse plants and animals that thrived in one landscape may increasingly end up living in a very different landscape because of climate change. These changes are already occurring.

Scientists have found that globally, species are moving an average of 11 miles north and 36 feet higher in elevation each decade to find more hospitable places to live.

How can conservationists continue to protect biodiversity if the cast of plant and animal species keeps moving to escape growing climate impacts? The answer, according to Dr. Mark Anderson, is to focus on the future. Anderson, director of science for TNC's Eastern US Division, for years has studied connections between climate change and shifting biodiversity. Anderson is also a GIS advocate who understands the crucial role that GIS plays for land trusts, state agencies, and other entities planning conservation projects. Conservationists have limited funding and face huge challenges in their work preserving biodiversity and clean water and protecting the health of landscapes. They need GIS to help identify and focus on high-priority projects.

Data from scientists, universities, and state and federal agencies provides valuable insight into current biodiversity, habitat, and other natural features. So Anderson wanted to see how GIS could identify places that would have long-term resilience as the climate changes.

Anderson and his team found that landscapes with diverse physical characteristics—such as steep slopes, tall mountains, deep ravines, and diverse soil types—create numerous microclimates that offer plants and animals the opportunity to move around their local neighborhood to find suitable habitat to escape rising temperatures, increased floods, or drought. Anderson and his colleague Melissa Clark mapped natural highways across the country—connecting corridors that allow species to move safely within and between these climate-resilient neighborhoods. While studies show that species are moving to cooler places, nearly 60 percent of US lands and waters are fragmented by human development, blocking species' movement, and preventing them from finding new homes.

"Nature is on the move," Anderson said. "It's not enough to have isolated and disconnected landscapes that are resilient to climate impacts. Species also need a way to reach these resilient sites. While some species will be able to relocate to new homes within their local resilient neighborhoods, others will need to move great distances to entirely new landscapes. If these pathways are destroyed, many species could disappear forever."

To create the Resilient and Connected Network mapping tool, Anderson and his team compressed 200 gigabytes of data into a useful tool that worked seamlessly, said Erik Martin, a TNC spatial ecologist who helped lead the GIS work. "The map reflects a ton of data, pulling from hundreds of datasets from all around the country."

Using Esri's advanced spatial analysis and modeling tools, TNC scientists Clark, Arlene Olivero Sheldon, and Analie Barnett developed, tested, and retested methods to map and measure characteristics of resilient lands. Their analytical and problem-solving skills, coupled with their determination and passion for conservation, took the idea of a network of resilient lands from concept to reality.

One of the big challenges, according to Anderson, was persisting through the 10-year lifespan of the project. During that time, the project grew organically region by region, starting with the northeastern United States. To ensure accuracy as the map was built, the team had reviews by some 270 scientists, who looked at geography at localized and nationwide scales, depending on their expertise. Aiding in that process was the ability to share regional versions via ArcGIS StoryMaps.

Many other Esri products and apps were essential for the job. These included ArcGIS Pro, ArcGIS Online for hosting tiled map services, ArcGIS Enterprise for hosting dynamic map services and geoprocessing services that run custom ArcPy-based scripts, and the ArcGIS API for JavaScript™ to develop the front end of the tool. All those components came together, making the map fast and responsive for users.

The team relied heavily on ArcMap and ArcGIS Pro, using 30-meter scale raster data of the entire country for analysis and mapping. The ArcGIS Spatial Analyst™ toolsets contained the team's go-to tools. Python and ModelBuilder™ were used to test analysis and iterate steps in each region. Getting feedback from regional scientists helped shape the final product. The team used the quick publishing capabilities of ArcGIS Online. StoryMaps stories were key in explaining analysis results and showcasing resilient places across the United States.

"Each of these components is like a building block of a house. We used the ArcGIS API for JavaScript to build the frame of the

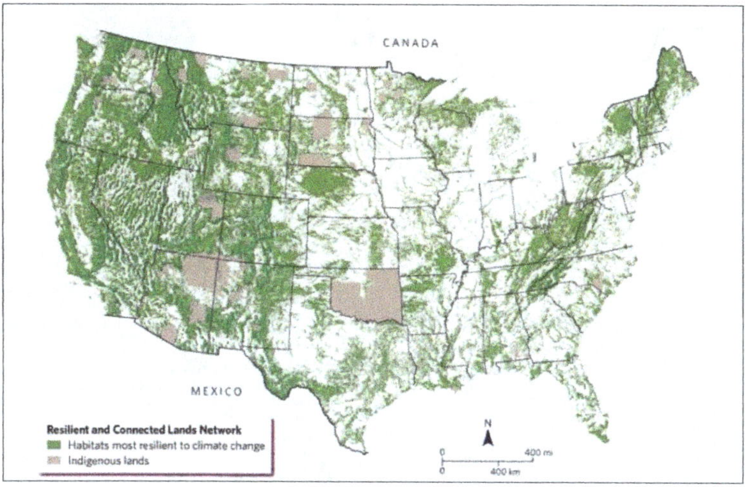

The Nature Conservancy scientists have mapped a network of landscapes across the United States with unique topographies, geologies, and other characteristics that can help withstand climate impacts.

house—the structure of the app. Then [we] connected all of the electrical and plumbing—the map and geoprocessing services that feed into the app," Martin said. "The Esri tools integrate all of the parts of this data into one useful app."

When the nationwide map was completed, one of the big surprises for Anderson was the large number of areas around the country that are resilient to climate change. "When I first envisioned this, I thought there would be a bunch of big, broad areas. It's really like a web, or a network of connected places."

The map includes these and other large and small resilient and connected lands:

- Nevada's Monsoon Passage, a natural highway of mountain ranges and wet valley bottoms that extends from Lake Mead through Great Basin National Park to the Idaho border.

- The Cumberland Forests area, which spans 253,000 acres across Tennessee, Kentucky, and Virginia, safeguarding wildlife habitat and storing millions of tons of carbon.
- Wisconsin's Kettle Moraine region, which was shaped by glaciers during the last ice age and contains diverse features such as large kettle lakes and 300-foot-high ridges.
- Bobcat Alley, a 32,000-acre forested corridor in northwestern New Jersey that provides habitat to state-endangered bobcats.

Along with providing safe places where species can thrive, the network of resilient lands also brings benefits to people. The lands mapped in the eastern United States, for example, contain 75 percent of the region's sources of drinking water, generate billions of dollars in outdoor recreation, store 3.9 billion tons of carbon, and mitigate 1.3 million tons of pollution.

"To achieve conservation at the scale needed, we must collaborate with people and organizations across both the public and private sectors," Anderson said. "Keeping these resilient areas safe and healthy will require a wide range of conservation practices including such things as sustainable management, public and private land acquisition, and easements."

Numerous state and federal agencies and land trusts have already incorporated the map's data into their conservation planning. Among the growing number of state agencies incorporating the map into their state Wildlife Action Plans are Connecticut, Massachusetts, New Hampshire, New York, and South Carolina.

The urgency of protecting lands and biodiversity given the challenges of climate change is clear and pressing, according to Andrew Bowman, president of the Land Trust Alliance. At Rally 2019: The National Land Conservation Conference, the country's largest gathering of land trusts, Bowman called for safeguarding biodiversity for

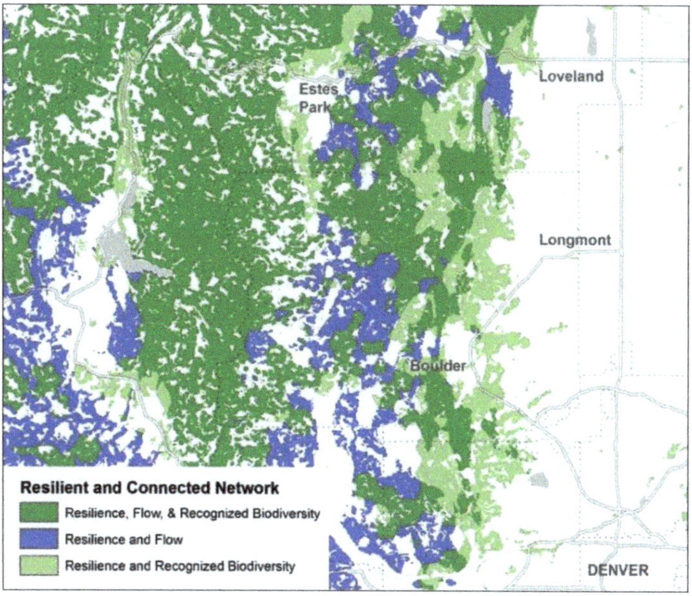

Among the things that make places resilient to climate change are diverse physical characteristics—such as steep slopes, tall mountains, deep ravines, and diverse soil types—that create numerous microclimates and allow plants and animals to move to suitable habitat. The landscapes also have good connectivity, allowing movement across local microclimates and between climate-resilient landscapes.

the future. The Resilient and Connected Network is an important tool for that urgent task.

"This map provides a new way to look at the landscape for purposes of prioritizing lands for the conservation of biodiversity," Bowman said. "I've watched this work take shape over the last decade, and I truly believe it marks a paradigm shift in how we think about large landscape conservation."

This story originally appeared as "Conserving a Network of Climate-Resilient Lands" by Eric Aldrich in the Winter 2021 issue of *ArcUser*. All images courtesy of The Nature Conservancy, unless otherwise noted.

IDENTIFYING VALUABLE PARCELS

Meghan T. Jones, LJB Inc.

THREE VALLEY CONSERVATION TRUST (TVCT), A NONPROFIT land conservation organization in Oxford, Ohio, needed to identify and prioritize land parcels for conservation easements that would help protect streams and riparian zones, promote prime production of agricultural land, and decrease the potential for habitat fragmentation.

Nonprofit land trust organizations, such as TVCT, implement environmental planning by working toward permanently preserving habitat, open space, and dynamic landscapes on private land, using different conservation strategies such as conservation easements. This strategy permanently limits land uses through a custom legal agreement to protect natural and agricultural values of the land. According to the Land Trust Alliance, in 2018 there were 1,363 land trusts in the United States.

Founded in 1993, TVCT primarily targets riparian habitat and agricultural land in southwest Ohio. It protects land through conservation easements that span seven counties in southern Ohio and Indiana. TVCT aims to implement widespread protection of open space, watersheds, and communities for present and future generations.

The land trust works toward this goal through sale or donation of conservation easements. Using federal and state grant funding, TVCT places permanent easements on private property. Easement acquisition is restricted by limited funding sources. Land managers create efficient conservation planning processes by designating land with high ecological value as high priority. Local land conservation organizations such as TVCT can benefit from prioritization tools that enable data-driven decision-making. TVCT funding comes from organizations and agencies such as The Nature Conservancy, the

Ohio Environmental Protection Agency (OEPA), and the Local Agricultural Easement Purchase Program (LAEPP).

To spend its available funds more effectively, TVCT wanted a parcel prioritization tool developed that would assist in targeting easement placements. A research assistantship was granted to Meghan Jones, a graduate student at Miami University, who created the tool for her master's thesis. Jones developed the Land Priority Protection Model (LPPM) using ModelBuilder, a visual programming language for building geoprocessing workflows in ArcGIS, to identify areas of land best suited for conservation. The tool applies conservation criteria to private land to weight, rank, and strategically score parcels.

TVCT identified three overarching evaluation criteria for identifying potential parcels of high value: presence of streams, adjacency to existing protection, and presence of desirable land cover/land use (land cover/land use types include wetland, forest, and agriculture use). Each criterion is evaluated within its own submodel and calculated individually. The final conservation model is computed last to accommodate all other criteria calculations. While the primary purpose of the model is to identify high-priority land to apply conservation easements, the model needed to be versatile so that variables can be excluded or weighted differently.

Three Valley Conservation Trust uses publicly available data sources such as the National Land Cover Database and the National Hydrography Dataset for analysis. The data was imported into ArcGIS and reprojected to NAD83 Ohio South State Plane Feet, clipped to the study area, and imported into a geodatabase before being included in the model.

LPPM automates the process of prioritizing parcels based on various conservation scenarios. The user determines whether to include or exclude criteria and assigns relative weights to each criterion. For example, if a funding agency prefers different conditions or criterion weight for conservation easement applications—such as the presence of wetlands—those variables can be weighted higher before running the final priority model.

Each criterion for identifying potential parcels of high value is handled by a submodel. Each submodel has a specific purpose: to create an attribute field for holding score. For example, the Presence of Streams submodel selects every parcel in the study area that intersects a stream and assigns it a score of 100 because a stream is present. All parcels that don't intersect a stream receive a score of 0, denoting the absence of streams. The appended parcel dataset is then used to run the final conservation model.

The final analysis adds a field to the parcels dataset to store the final priority score. This score is calculated as a sum of each criterion present for each parcel multiplied by the individual weight assigned, which ranges from 0 to 100. A weight of 100 represents the most desirable land for conservation, and a weight of 0 represents the least desirable land for conservation. Results are symbolized in even quantiles from 0 to100 based on the final conservation index parcel score. Of 201,477 parcels evaluated, only 514 (0.2 percent) scored a perfect 100, while 3,777 parcels (1.87 percent) contained the highest priority score ranging from 75 to100.

Using LPPM, TVCT can analyze open-source data and identify at-risk ecosystems. Mapping natural resources using weighted criteria helps illustrate how private parcels of land impact biodiversity and ecosystem services on a landscape level. For example, this model could provide the land trust with a way to examine which parcels should be set aside to provide contiguous habitat along a riparian corridor, rather than focusing on land use or land cover.

Conservation funding is limited, and so is the amount of land the funding can ultimately protect. TVCT scrutinizes available funding

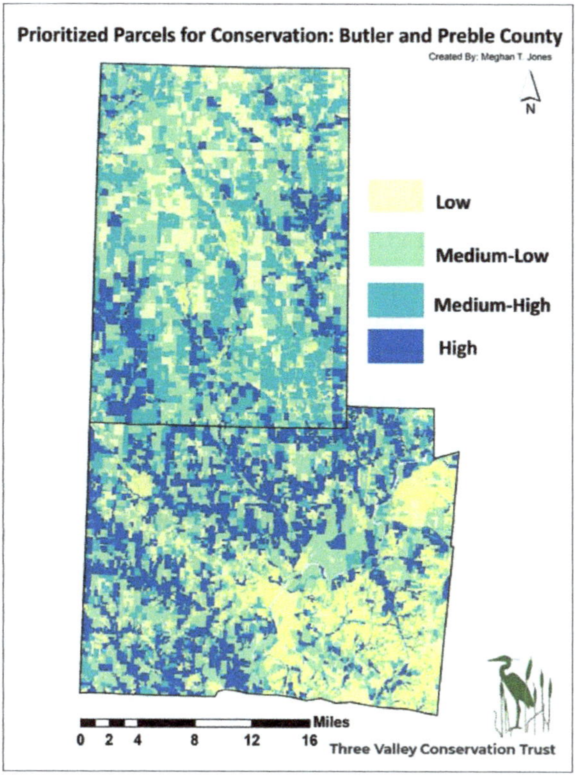

The Three Valley Conservation Trust created the Land Priority Protection Model to prioritize parcels for conservation in southwest Ohio.

for projects and parcels that will provide the greatest ecological benefit. LPPM allows TVCT to be proactive and educate landowners who may be less inclined to support local land protection on the ecological value of their property and increase community involvement with habitat preservation or protection. LLPM helps communicate land protection issues as they relate to landowners' holdings and target donors for restoration projects.

After the initial analysis and selection of strategic habitat using the LPPM, TVCT exported the selected parcels to remotely piloted aerial systems mapping software loaded on Apple iPad devices. The staff used this technology and software to obtain current aerial images and vegetation indexes. In addition, field staff used the maps to guide biological surveys, draft habitat restoration and protection plans, and create detailed presentations for meetings and workshops.

Going forward, TVCT continues employing LPPM to identify habitat zones containing vital or at-risk native flora and fauna. As funding sources are identified, the conservation trust adjusts data layers and weights to align strategic mapping with funding partner criteria, which allows TVCT to maximize the impact of collaborative funding from multiple sources; visually communicate conservation initiative goals, planning, and assessment steps; and significantly increase community outreach and financial support.

This story originally appeared as "Identifying the Most Valuable Parcels to Protect" by Mehgan T. Jones in the Fall 2019 issue of *ArcUser*. All images courtesy of the Three Valley Conservation Trust, unless otherwise noted.

TAKEAWAYS

CLIMATE CHANGE AND THE DIMINISHMENT OF VIABLE ecosystems are increasing concerns for communities and organizations tasked with maintaining shorelines, protecting critical habitat, and preserving natural landscapes. GIS mapping and spatial analysis are critical tools for predicting environmental impacts, planning remediation and conservation, and gaining political and public support for projects related to climate change and conservation.

- **Prioritize land for conservation:** GIS helps land administrators accurately identify and prioritize land parcels for conservation easements that protect streams and riparian zones, promote prime production of agricultural land, decrease the potential for habitat fragmentation, and secure federal and state funding.

- **Mitigate flooding and shoreline loss:** GIS mapping and analysis provides a science-based framework for creating adaptative strategies when responding to climate-related events that impact the natural processes of watersheds, creeks, and shorelines. GIS tools help organizations understand and continuously measure the chemical, physical, and biological health of wetlands, estuaries, and river systems.

- **Promote sustainable development:** Using GIS mobile applications to collect field data and then using that data to make and share maps with local communities helps governments and nonprofits build awareness of the importance of sustainable development. Accurate mapping shapes people's thinking about preserving biodiversity and protecting threatened ecosystems and can directly reduce poverty by creating jobs in ecotourism and agriculture.

- **Identify valuable parcels:** Organizations use GIS to visualize and organize their strategies for preserving habitat and open space. Different conservation tactics, such as conservation easements, help provide the greatest ecological benefit. Mapping and geographic analysis can educate landowners and help grow support for local land conservation and increase community involvement in habitat preservation or protection.

PART 4

EQUITY AND SOCIAL JUSTICE

LAND IS PRECIOUS. FOR MANY PEOPLE, PROPERTY ownership represents a step up in economic well-being and a measure of pride. For others, property embodies social status, power, and control. However, land is so much more than a physical possession—it is the perception and reality of place that defines who we are, our history, and our struggle to overcome adversity and claim our place in the world.

Historically, as land was (and continues to be) divided and parsed into places of specific purpose, entire cultures of indigenous people were displaced—pushed out of the way in the name of progress. In other cases, displacement is the result of ethnic prejudice, political maleficence, war, catastrophe, and blatant greed, resulting in refugee camps, reservations, blighted communities, and other forms of social injustice. Even now, world governments, including the most prosperous, face serious matters of land restitution and rights, affordable housing, and the loss of cultural heritage.

With GIS, land administration agencies and nonprofit organizations use location intelligence to better understand the social costs of land inequity. GIS helps them find new ways to address impoverished communities and empower displaced people with scientific tools to gain a stronger voice in land use and management. Geospatial data and analysis helps identify where and why land administration decisions negatively impact people who are already dislodged from the decision-making process and whose lives and cultures are diminished or ignored when the lands of their people are repurposed or given away.

Real-life stories

The real-life stories in this section represent but a few examples of how some organizations are delving into land equity and social justice issues, using GIS to investigate, document, and implement solutions. GIS allows these organizations to expertly explore the geographic and social characteristics of place and uncover the history of land and property, including the cultural and emotional significance of changing landscapes. A common thread throughout these and other examples is the importance of using precise mapping and spatial analysis to define important places, while recognizing the conflicts and opportunities that change land administration practices. In every case, GIS also is used for communication and collaboration, providing a visual language for discussing and sharing land equity and social justice information.

EMPOWERING WOMEN LANDOWNERS

Jen Van Deusen, Esri

ACCESS TO ARABLE LAND HAS BEEN CENTRAL TO HUMAN life for at least 12,000 years. Yet even today, ownership of rich farm soil has largely been denied to one gender. Women, who make up half of the global adult population and 43 percent of the agricultural workforce, account for less than 15 percent of farmland owners. Nearly 40 percent of the world's economies still limit women's property rights, according to the World Bank. In the remaining 60 percent, male-centric cultural norms and legal inefficiencies still hinder women's rights.

Ensuring women have the right to own land could lift millions of people out of hunger, reduce rural poverty, and improve sustainable natural resource management, according to studies from the United Nations (UN), the World Bank, the women's rights nonprofit Landesa, and others. Such sources also agree that achieving global gender equity will start with local efforts—often by putting more information in the hands of women.

"It's about data empowerment," said Amy Coughenour Betancourt, CEO of Cadasta, a global nonprofit that connects communities with technologies for recording land and resource rights. "We approach it as enabling local stakeholders to document, secure, and manage their land rights and helping communities understand and use the data that's collected."

Giving women the choice to take ownership of their land starts with building a better understanding of their legal rights, and that begins with capturing digital records. Better access to maps and property data can also simplify proof of ownership, which can be expensive and time-consuming to obtain, particularly in rural communities. In Mozambique, for example, Cadasta worked with local leaders and the National Cooperative Business Association Cooperative League of the United States of America (NCBA CLUSA) to make it easier for local farmers to get a certificate of landownership. Digital maps created with GIS technology eliminated several surveying steps, lowering time and costs from upward of a year and 400 US dollars in some cases to as little as three months and $30.

"Mozambique is a really good example of how the technology facilitates these processes," Coughenour Betancourt said. "And the efforts in Mozambique emphasize women-led farms. Women are given priority, and if the woman is married, she's included on the certificate along with her husband." Coughenour Betancourt notes the importance of creating formal records and getting women's names documented in landownership. When they aren't included in writing, women are particularly vulnerable to losing their homes or livelihood.

Within the last decade, the UN has identified land and housing security, the role of women in agriculture, and women's land rights as key indicators in meeting its Sustainable Development Goals (SDGs) around poverty, hunger, and gender equality. Most of the world's poor and hungry are women and girls. Focusing on women's rights targets an especially vulnerable population.

Evidence also shows that the benefits of gender equity in land and resource distribution extend far beyond women's empowerment, uplifting not just women but entire communities. Women lose an estimated 20 to 30 percent of farming capacity because of limitations placed on their access to land, tools, and education. With equal access, women could produce enough food to lift between 100 and 150 million people out of hunger.

In addition, Coughenour Betancourt said, "Women inherently invest more in their families with their income," referencing a collection of studies from around the world that show improved outcomes in health, food access, and education when women are included in family landownership. Women farmers are also more likely to grow crops that can feed their families and support local commerce.

At a time when COVID-19 has disrupted supply chains and thrust millions worldwide into food insecurity, leaders are looking for ways to bolster sustainable food systems that keep communities fed when global distribution fails. Evidence shows that women's access to land and education can help fill the gaps by boosting local food production and growing nutritious, resource-efficient crops.

"A sustainable food system has many essential components," Coughenour Betancourt said. "Three of which are land rights, climate-smart agricultural practices, and women's empowerment. And one of the best ways to promote women's empowerment is to give a woman a right to land."

Land to farm has long provided for basic needs such as food, shelter, and income. Beyond survival, though, land rights can be a gateway to a person's financial independence, social mobility, and political inclusion. This cycle of opportunity is central to the sustainable future the UN SDGs enable. Transparent, accessible, and equitable systems empower more people, and in turn, create stronger communities better equipped to withstand the challenges of our time.

This story originally appeared as "If More Women Owned Land, More People Might Be Fed" by Jen Van Deusen in the *Esri Blog*, August 10, 2021.

DEFINING LAND RESTITUTION

Jim Baumann, Esri

INEQUALITY IN LANDOWNERSHIP HAS PERSISTED IN Colombia since the country won its independence from the Spanish Empire in 1819. To pay for the massive war debt it incurred, the newly sovereign nation quickly sold off vast tracts of poorly documented public land, which often ended up in the possession of already wealthy landowners.

During the mid-1960s, guerrilla groups formed—most notably the Revolutionary Armed Forces of Colombia (known by its Spanish acronym FARC). These groups existed to promote greater equality throughout the country, but they fast became a polarizing force. This exacerbated political and social problems throughout Colombia, particularly in the rural areas, where guerrilla or opposition paramilitary groups often controlled villages and their surrounding lands after forcibly displacing the original landowners. Over the course of its six-decade duration, the Colombian Conflict resulted in 260,000 deaths and the displacement of 6.9 million people.

In 2010, former Colombian president Juan Manuel Santos announced in his inauguration speech his intention to return the land to those who were displaced by the Colombian Conflict. “It is not only an inescapable historical debt, but it is also a first step towards the construction of peace in the rural areas of the country,” Santos said.

The amount of disputed land totals 7 million hectares (17.3 million acres). This includes private properties and vacant and unclaimed land.

The following year, Congress approved Law 1448, the Victims and Land Restitution Law, which specifies the legal procedure for restoring lost land to victims who were forcibly removed from their properties. The procedure consists of two parts: one is administrative, which includes formally registering the land claim, and the other is a judicial appeal for restitution.

"The Land Restitution Law is one of the most important acts of the Colombian government in providing a lasting solution to the humanitarian crisis precipitated by the decades-long conflict," said Ricardo Sabogal, former director of Unidad de Restitución de Tierras (URT), Colombia's land restitution office. "It is an effective administrative and judicial mechanism that resolves, in a just manner, the

Using GIS helps Unidad de Restitución de Tierras, Columbia's land restitution office, establish where land claims overlap with specific microzones (green), ethnic areas (red), and more than one municipality or department (orange).

complicated issues for the millions that were displaced from their land. It also encourages lasting peace and a consolidation of democracy in our country. The process consists of both an administrative step to formally register the land claim by the displaced farmer and a judicial action to resolve the land restitution claim made by both parties. In this sense, it is a combined process because it includes involvement by both the country's executive and judicial branches."

The URT performs the administrative procedures. The office receives, reviews, and decides for all property claim applications. The National Land Agency then officially records the disputed land, which is an essential legal step in the restoration process.

"Because of the well-known informality regarding land possession in Colombia's rural regions, Law 1448 also specifies that the URT must physically and legally define those properties that did not previously have cadastral or registry information associated with them so that they could be included in the records of the Oficina de Registro de Instrumentos Públicos (ORIP), the nation's land registry authority," said Sabogal.

Complying with this requirement was made possible through GIS technology, which helped the URT collect and analyze data that identifies distinct properties and specifies their precise location and size.

"Law 1448 stipulates that in order to collect information for restitution requests, it is necessary to first examine the available institutional information, such as ownership records, for the physical identification of the properties," Sabogal continued. "Unfortunately, the rural cadastral property information is not the most reliable. According to the Consejo Nacional de Política Económica y Social [National Council of Economic and Social Policy], in 2016, 28 percent of the national territory did not have cadastral records, and 63.9 percent had outdated cadastres, which totaled 722 municipalities.

Also, among those 187 municipalities in the country identified as having a high incidence of armed conflict, 79 percent did not have basic cadastral information. Moreover, only 41 percent of the national territory was mapped with appropriate basic cartography at a scale needed for technical cadastral work. and as a rule, the cadastre only identifies people who had some sort of document indicating an affiliation with a property, which excluded most land claimants."

To establish landownership, the URT collects data in the field and gathers historic information concerning each property's years of abandonment or dispossession. Staff from the URT also tour the boundaries of the property along with the applicants and their neighbors to determine accurate dimensions, boundaries, terrain details, and possible conflicts that are added to the evidence presented to the restitution judges. The URT uses satellite imagery and aerial photographs to conduct cartographic and cadastral archaeology, which helps the office establish property boundaries at submeter precision.

For applicants going through the administrative and judicial processes of recording a land claim, the use of GIS technology has played a significant role in validating the size, location, and ownership of each property.

"Technical and cadastral professionals from the URT, with assistance from Esri Colombia, are using ArcGIS to design methodologies to capture, structure, define, analyze, and exchange geographic information to determine the correct identification of an applicant's land," said Helena Gutiérrez, president of Esri Colombia SAS, Esri's official distributor in Colombia. "The process involves the creation of a sequence of versions of the property as the investigation progresses. The process of this technical adjustment of the mapped polygon representing the property is called its 'cadastral life.'"

"The URT has created a data portal using ArcGIS Online to provide easy access to geospatial services in support [of] this work," said

Jorge Bonil, the URT's cadastral technical director. "In addition, we used ModelBuilder to create an automated reporting application that accesses and processes massive amounts of information from multiple sources and performs various analyses on that data. The tool identifies conflicting claims for the same land, claims in protected areas, unusual economic activities related to a particular piece of land, identification of the licensing or construction of infrastructure and energy production, and so on. Reports from these analyses are then generated for the administrative and judicial procedures."

The land restitution initiative also uses thematic cartographic data produced and maintained by different Colombian governmental agencies. This collection includes data on national protected areas; mining and hydrocarbon extraction; road and energy infrastructure construction; strategic conservation, such as wetlands of international importance; government land usage and planning; land mines and unexploded ammunitions; and proprietary information prepared by the URT, such as applications for micro- and macrozones.

In addition, the URT employs satellite imagery and aerial photography to determine access routes to specified properties and the physical characteristics of the landscape and topography. For land disputes, they are sometimes used to divide parcels and determine equivalent land assets. Imagery analysis also helps the URT ascertain the status of collective territories and current land use.

"The restitution laws have set an important precedent for the availability and exchange of information among the government agencies," Gutiérrez pointed out. "It has changed the institutional perspective towards sharing information by highlighting common objectives, which strengthens our government."

Previously, she said, accessing government data was complex from a technological and legal standpoint.

"As part of the land restitution program, the URT participated

The URT starts the land claim process by getting information from the applicant to make a preliminary determination of the location of the land and then combining that information with official government data. That process associates the land with URT-generated microzones and sets in motion the administrative process.

in the development and testing of the LADM-COL model for land administration," explained Gutiérrez, referring to a version of the international Land Administration Domain Model (LADM) standard that was adapted for Colombia. "This has allowed us to test and adapt those workflows needed for the supply and delivery of cadastre information within the established standards that should be included in the procedures implemented by the government."

"The application of GIS technology to the process of land restitution assures that the land restored to its owner has been accurately identified and documented for legal inclusion in the land registry database," added Sabogal. "GIS also helps with identifying the ways the property is related to nearby land to determine potential environmental implications, mining and energy production, and infrastructure policies—thus guaranteeing a proper use of the land without impacting the regulatory frameworks of other policies. It has also allowed us to perform geostatistical modeling to determine trends and explore scenarios so that we can develop land administration polices using more accurate data."

The second, judicial stage of the process is carried out by judges who specialize in land restitution proceedings. Judicial authorities decide if the claim is legally and materially admissible, in which case the property is included in the Superintendence of Notaries and Registries. In cases where restitution of the property is impossible, the judges have the authority to order compensation in favor of the victim or in favor of third parties, as long as their actions were done "in good faith." The government processed 1.5 million hectares of land as of 2019, with a goal of eventually processing 100 percent of the land restitution claims.

The URT also has an important role in these judicial proceedings because it provides free legal representation to victims who request it. As part of this service, the office created a Mobile Victim Attention and Orientation Unit. Representatives from the National Victims Unit, the Public Defender's office, and the Ministry of Justice drive a custom-made vehicle into rural municipalities to provide legal aid to people who have been displaced. This ensures that their interests are being respected, as specified in the law.

The land restitution process in Colombia not only provides legal services to thousands of displaced people, but landownership also allows them access to credit that can help them further develop their land. In addition, as landowners, they pay taxes on their property, which helps improve government budgets.

"The speed in applying Law 1448 to the land restitution process has gained a level of success that has never been achieved by any other judicial mechanism in the country," concluded Sabogal. "This has been made possible due to the foresight and determination of our government and its implementation by various agencies. It is important to remember that some of this work was carried out in the middle of the armed conflict, which created a unique series of complexities and difficulties in its implementation."

This story originally appeared as "In Colombia, GIS Defines Land Restitution for Forcibly Displaced Persons" in the Spring 2019 issue of *ArcNews*.

SUPPORTING LAND RIGHTS

Brent Jones, Esri

IN INDIA, HAVING *PATTA* (OR PROPERTY TITLE) FORMS THE basis for social and financial opportunity. It often provides the only acceptable proof of address required to access government benefits, enroll children in school, and open a bank account. A program in India's state of Odisha has begun to deliver land rights to people living in lower-income areas.

"This project was conceived from the realization that slum dwellers are the lifeline of the city and our state can't be developed without uplifting them," said G. Mathi Vathanan, commissioner-cum-secretary, Housing and Urban Development Department, Government of Odisha.

The Tata Trusts began the effort with pilot studies that gained momentum with the historic passage of the Odisha Land Rights to Slum Dwellers Act in August 2017. The legislation puts a program in motion to identify, map, and issue title for parcels of land in 2,000 areas that house a population of 1 million people.

"This landmark legislation is going to change the face of slums across the country," said R. Venkataramanan, managing trustee, the Tata Trusts. "Hopefully, this effort in the slums of Odisha will have a ripple effect across the country with a positive social and economic impact."

"This first-of-its-kind program has shown the potential of bottom-up data documentation leading to formal certificates of occupancy in urban informal communities," said Frank Pichel, chief programs officer at Cadasta Foundation, a nonprofit that provides technical tools and services to support the efficient documentation of land rights globally.

The legislation promises a Certificate of Occupancy of 30 square meters (roughly 320 square feet) free of cost to economically and socially disadvantaged people, and financial assistance of up to 200,000 rupees ($2,900) that the homeowner can use to create a permanent dwelling. Odisha's effort is part of Prime Minister Narendra Modi's larger Housing for All Mission that aims to build 20 million urban housing units and 30 million rural homes by 2022.

Worldwide, 70 percent of the population live on land without holding title. Peruvian economist Hernando de Soto helped raise awareness about the impact of granting land title by writing several best-selling books. He argues that guaranteeing urban property rights is a precondition for alleviating poverty.

Many governments have experimented with de Soto's ideas, and the World Bank has loaned funds to support related projects; however, the complexities of land reform have yielded several false starts.

According to Transparency International, the land sector is the third most corrupt government sector globally behind police and judiciary.

"A lot of places where we've worked, the slums are owned by the politically or economically powerful," Pichel said. "They are happy for it to remain informal, because they can control it and continue charging unreasonable rents knowing they can always evict a person if they don't pay."

Even where land records exist, historical information can be inaccurate or not up to date. Often, when reforming land records, it's cheaper and easier to inventory existing occupancy rather than correct old records.

"In other countries, we've seen deed books with blank or torn-out pages," Pichel said. "A torn-out page means someone's deed disappeared. A blank page means the official can predate a deed to give favor to the person that pays them the most."

Streamlined digital workflows have started to reverse these issues.

"Technology brings in ease of doing this accurately, transparency in the process, removes discrepancies, reduces the dependency on human interventions, and brings in speed in execution," Vathanan said. "It also helped avoid large-scale disputes, litigation, and discontentment that would have resulted in frustration and failure."

The government's partner, the Tata Trusts, brought in global best practices by mobilizing partners like Norman Foster Foundation, Omidyar Network, and Cadasta. The government also partnered with more than 27 local civil society organizations for data collection and validation.

They started data collection by using drones to capture high-resolution imagery of the slum areas. This imagery formed a backdrop within the GIS for drawing boundaries and assigning a plot number. Next, data collectors went door to door using a mobile application to collect survey information from each household.

"We brought in state-of-the-art technology and expertise to complete the task in the most efficient manner," said Shishir Ranjan Dash, who leads the project for the Tata Trusts.

Each step of the way, the teams relied on Esri technology, tailored by Cadasta, to capture land information. Workflows include handheld applications configured for the survey task, a claim review process, steps to transfer the formalized land records to the government, and the delivery of four formal certificates of occupancy to the occupants.

Using GIS apps, the field teams, made up of one field manager and five data collectors per neighborhood, steadily captured records for 200–250 households per day. This fast pace is important given the extent of the task. The data collectors are local residents who capitalize on their local knowledge to speed the work and ensure its legitimacy.

Smartphones are central to the process. Field crews use phones for all steps, and the popular texting application WhatsApp is used to communicate progress.

"The number of people in emerging economies interacting with GIS via a web browser on their smartphone as opposed to a computer is incredibly high," Pichel said. "A few years ago, I would get incredibly frustrated by bandwidth constraints, but that's no longer the case."

"This program was deliberately designed for doorstep delivery of services," Vathanan said. "We are pleased that land certificates have been distributed without requiring the slum dwellers to visit any government office even once."

To date, more than 24,000 certificates have been distributed to slum dwellers and an additional 50,000 will be delivered by the end of March 2019.

In other countries that have implemented land title reform in informal settlements, long-term stability has evaded residents. From Cambodia to the Democratic Republic of the Congo, giving land title in lower-income areas close to commercial centers has resulted in quick land sales and a total turnover of people, with the original settlers displaced within a year's time.

"In the case of Odisha, the property documentation that residents get enables the properties to be inheritable, but nontransferable," Pichel said. "So, they can't sell it, but they could leave it to heirs, and it's mortgageable."

Studies show that land title helps lift residents out of poverty by erasing the fear of eviction and cultivating safety and dignity. Title to a house turns it into an insurance and savings tool to provide security for owners during old age or bad times.

"What you often find in slums where there is a risk of eviction is that somebody remains in the house at all times," Pichel said. "That way, they can move out all the possessions quickly. With someone needing to be there, it means a child's not going to school or an adult isn't out earning an income."

Land title also shifts investments into more permanent structures. Instead of spending money to rebuild every year after monsoon storms hit, the resources can be placed into more permanent structures that withstand wind and rain.

"It is important to create a sense of ownership among the people," said Shikha Srivastava, the Urban Habitat Portfolio lead at the Tata Trusts. "To ensure this, we are upgrading the slums through a participatory decision-making process where the community shares their development vision."

The State of Odisha's investment includes roads, drainage, fresh water supply, toilets and sewers, streetlights, common work sheds, parks, and playgrounds. This expanded vision of the program goes beyond the original promise of land title and includes the transformation of slums into livable habitat.

This story originally appeared as "How One Million People in India's Odisha Slums Gain Land Rights" by Jim Baumann in the *Esri Blog*, February 11, 2019.

AIDING AFFORDABLE HOUSING

Rob Matthews, Esri

IN 2017, RESIDENTS AND FAMILIES LIVING IN A 33-UNIT mobile home park in Chapel Hill, North Carolina, received news from the property owner that within a year they would have to relocate so the property could be redeveloped. After the tight-knit community pleaded for help, the Chapel Hill Town Council mobilized a small team of town employees to explore affordable housing options.

"About 90 percent of the residents in our mobile home communities are Latino, primarily Spanish speaking, from low-income households. Many residents are long-standing members of our community," explained Sarah Viñas, assistant director, Office for Housing and Community, Town of Chapel Hill. "The town is really committed to have housing options available to these and all of our residents."

Going back several decades, the Town of Chapel Hill deployed a land-use strategy mandating specific areas for urban development and others for preservation. The rural buffer prevents development on town outskirts and limits growth, which in turn reduces the amount of available land and increases land costs—presenting a challenge for the town's Office for Housing and Community.

"The initiative to identify opportunities for affordable housing was uncharted territory for the town's GIS team. There are only four mobile home communities in town and they're all in relatively high-value commercial areas," said Dave Almond, senior GIS analyst in the Department of Technology Solutions, Town of Chapel Hill.

Town staff began by reaching out to the mobile home park residents at meetings and conducting surveys to learn more about who they are, their household demographics, and their housing preferences. The desire was to craft solutions based on specific needs. After

Manufactured Home Communities

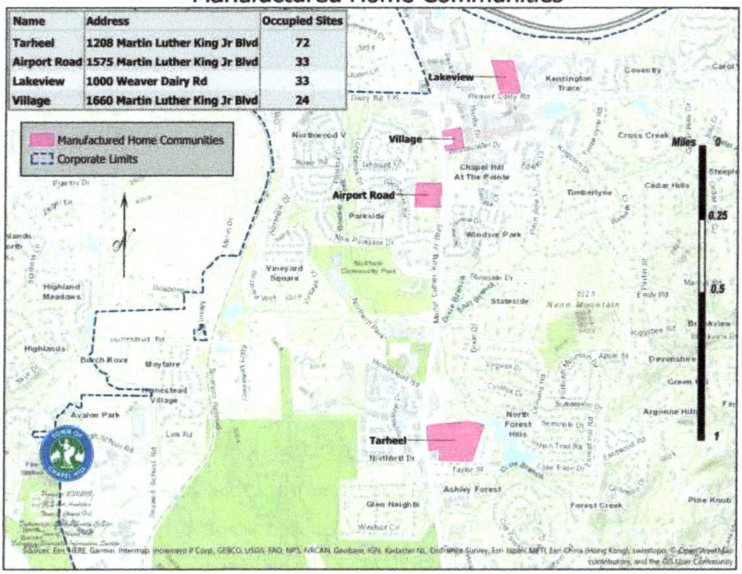

Name	Address	Occupied Sites
Tarheel	1208 Martin Luther King Jr Blvd	72
Airport Road	1575 Martin Luther King Jr Blvd	33
Lakeview	1000 Weaver Dairy Rd	33
Village	1660 Martin Luther King Jr Blvd	24

This map displays the location of all the manufactured home communities in the town of Chapel Hill.

Viñas and her team connected with all the mobile home park residents in Chapel Hill, they realized that a mobile home was one of their least preferred housing types for relocation. "We've been committed to the resident engagement piece, starting with what residents say they want and designing strategies around it," Viñas said. "As a town, we've been intentional about our approach to listen and get feedback so we don't create solutions that, in the end, may or may not work for people."

Viñas' department then worked collaboratively with Almond, who analyzed and mapped available town-owned land for relocation using the town's GIS. Almond created a comprehensive visualization of properties, including amenities near potential sites. After Almond did an initial series of suitability analyses for candidate properties, he created high-quality 3D scenes, derived from 2D operational data, and shared them online to gather staff feedback.

The immersive 3D maps helped staff visualize properties and possibilities. "I wanted to go as far as I could to accurately represent in the 3D space absolute constraints to development. Deriving site characteristics such as slopes, existing structures, and vegetation from high-accuracy lidar data allowed us to create spatially accurate 3D scenarios," Almond explains.

The 3D site models helped the team collect design criteria from technical reviewers including engineers, transit and public safety officials, and town planners. Each reviewer could comment on potential sites in terms of deficiencies or specific needs for regulation and development, enabling Almond to refine the 3D concept scenarios.

For Viñas, a data visualization provided some surprising insights, including the integration of affordable housing throughout the town. "Our affordable housing is spread throughout the community, and not concentrating low-income people in specific neighborhoods is really best practice in the field," Viñas says. "So, seeing the data mapped was really striking. There's a lot of benefits for neighborhoods to be diverse in terms of income, race, and other demographic factors."

For residents facing relocation, maps showed amenities located near potential sites as well as upcoming and recent developments near the mobile home park. Viñas says the residents "really reacted to that," as they hadn't understood all the development unfolding around them. "We weren't quite sure when we put that together if people would make the connection and if it would be clear," Viñas says. "I think it was pretty neat to see the real-life use of some of these mapping tools, which sometimes can feel difficult to understand."

According to Viñas, the maps created by Almond and the GIS team are essential for the Chapel Hill Town Council. Quarterly reports on affordable housing presented to the council include a heat map of affordable housing in Chapel Hill. "The visualization tools were really helpful for an understanding of the broader context of the issue," Viñas says.

While the developer poised to redevelop the mobile home community that triggered the investigation later withdrew the proposal, the mapping explorations spurred by this activity have informed the town's efforts to identify suitable sites for affordable housing development. Most Chapel Hill residents (72 percent) approved an Affordable Housing Bond Referendum in November 2018 that included $10 million to help support the development of 400 new affordable housing units and preservation of 300 existing affordable units over the next five years. The bond will fund and support several large-scale affordable housing projects, including public housing redevelopment and development on town-owned parcels.

"Initially, the interest was a public intervention for this specific community, looking for just one property that could house the 33 units needed," Almond says. "The bond provides the town with the resources to pursue potential affordable housing development that would serve mobile home park and other low-income residents on town land that the Town Council has prioritized for this purpose. So, I didn't expect the outcome; it was a wonderful surprise for our team and the deserving residents."

This bond initiative sparked more than just discussion of the future of affordable housing in Chapel Hill; it inspired renewed focus among town leaders. The Office for Housing and Community is working on a partnership with the University of North Carolina at Chapel Hill on the Building Integrated Communities (BIC) project.

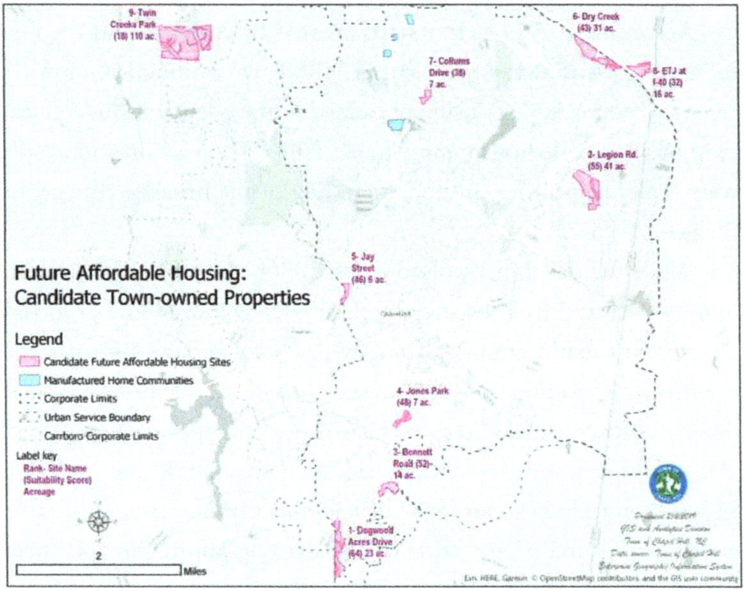

This map depicts the town-owned land being considered for future affordable housing development.

BIC is a multiyear community planning process that helps local governments in North Carolina engage with local foreign-born, refugee, and Hispanic and Latino residents to improve relationships, enhance communication, and promote civic participation and leadership. One aspect of BIC is looking at mobile home parks in and near Chapel Hill to see how far residents live from public transportation that they rely on to travel to and from work, school, and other critical services.

"I've been working on affordable housing at various levels, and this helps complete the picture for me," Almond explains. "Now, it expands to the outside of town and incorporates transit." Almond says he's gained a solid understanding of affordable housing, and as a GIS analyst, he wants to use GIS technology and analytics to help arrive at practical solutions that help people in Chapel Hill, a town with a population of about 65,000 people. "There's under 700 employees for the town government, and I'm one of three GIS analysts," Almond says. "So, we really have to champion the analytical capabilities of GIS because most people may not understand what's possible."

This story originally appeared as "Maps Aid Affordable Housing Push in Chapel Hill, North Carolina" by Rob Matthews in the *Esri Blog*, February 5, 2019. All images courtesy of the Town of Chapel Hill, unless otherwise noted.

TAKEAWAYS

GOVERNMENTS, NONPROFIT ORGANIZATIONS, TRIBES, AND other communities of indigenous people use GIS mapping and spatial analysis to better understand land equity and social justice issues impacted by land administration decisions and plans.

- **Empower women landowners:** GIS allows users to digitize land records; view property parcels on a map; and more easily find information such as land use, legal ownership, and value. Studies show improved outcomes in health, food access, and education when women are included in landownership, uplifting not just women but entire communities.
- **Overcome land dispossession:** GIS strengthens the gathering of historic information concerning property abandonment or dispossession, giving displaced people and marginalized cultures the tools to determine accurate dimensions, boundaries, terrain details, and conflicts over land rights, which in turn provides evidence for restitution judges.
- **Guarantee urban property rights:** GIS maps and mobile data collection applications support efforts to document landownership and alleviate poverty. Studies show that land title mapping and analysis reinforce efforts to lift residents out of poverty by erasing the fear of eviction, cultivating safety and dignity, and providing long-term security for owners.
- **Plan for affordable housing:** GIS allows planners and community leaders to visualize and gain new insights about affordable housing opportunities, while ensuring diversity in

terms of income, race, and other demographic factors, and helping people understand what is possible.

Learn more about applying GIS to land administration equity and social justice in the last section of this book, "Next steps."

NEXT STEPS

The geographic approach to land administration

USING GIS AS AN ESSENTIAL TOOL, LAND ADMINISTRATION agencies, local governments, and real estate companies can build a strong sense of location intelligence throughout their organizations, improve communication, and make working together easier and mutually beneficial. The stories presented in this book showed how land administrators, planners, consultants, and nonprofits apply GIS to spatially enable land data and applications. As a next step, readers can explore different ways to apply GIS to their organizations and, ultimately, experience GIS technology firsthand.

The information presented next is organized around the topics covered in this book and represents some of the most common strategies that land administration agencies and land-use professionals use to apply GIS to their organizations.

Parcels and property

Land administration processes and technology have continually advanced to move beyond paper-based systems and streamline outdated techniques for maintaining land records. But governments still face problems with land data aggregation and providing timely access to parcel and property information. With GIS, the work of land administrators can become more accessible, easier to understand, and thus more useful to more people. Community planners, assessors, real estate professionals, and developers can benefit from the same information when the data is based on geographic principles and spatial intelligence. As shown in the real-life stories in this book, some organizations are well ahead of the curve in terms of sophistication and innovation, while others are just beginning to move toward a geographic approach to land administration. The next checklist represents a few of the common steps taken to achieve consistent and reliable parcel and property administration.

Parcels and property checklist

- ☐ Gather staff to manage and maintain parcel and property information, such as land records, surveys, and urban and regional planning. Empower building inspectors and realty specialists to introduce the GIS-based parcel management system initiative. Appoint leads who will drive the initiative and an executive sponsor who will champion the initiative to leadership and across the organization.
- ☐ Produce a more accurate and useful parcel inventory by converting land record data into a GIS map layer or ArcGIS Parcel Fabric.
- ☐ Connect the GIS parcel map or parcel fabric with other land record data, legal descriptions, historical records, surveys, and related imagery and photos.
- ☐ Use GIS analysis to identify and locate parcels with missing or incorrect information, and update the information.
- ☐ Use GIS editing tools to compare and update parcel boundaries based on trusted surveys and legal descriptions.

- ☐ Compare updated GIS parcel layer or parcel fabric with existing zoning and other land-use boundaries, and work with planners to synchronize the parcel data, zoning, and land-use data.
- ☐ Share the new GIS parcel layer or parcel fabric with other departments, such as planning, engineering, police, fire, and public works.

Land management

Land managers focus primarily on land use to ensure that appropriate use of land is assigned to a parcel or area, the use is compatible with urban and regional planning objectives, and the designated use meets local, state, and national regulations and guidelines. Some countries lack a unified land management system, and in some cases, the system they have is not sufficiently implemented to serve as a general guide for land-use decisions. Rigorous land management practices are essential to ensuring local and national prosperity. These practices also mitigate conflicts between landowners, government agencies, and nonprofit land preservation organizations that can arise over infrastructure development such as highways and other large capital improvement projects. Land managers can use GIS to create systematic and unified land databases and analyze land-use data to discover problem areas, inconsistencies, and potential conflicts. The next checklist represents several proven actions that land managers use to monitor and update land management data.

Land management checklist

- ☐ Create a land use classification methodology using proven best practices, including GIS, to manage land-use data, visualization, and analysis.
- ☐ Use GIS analysis tools to discover current patterns of land use across jurisdictions based on national, regional, and local government agencies, specifically looking for gaps and inconsistencies.
- ☐ Combine GIS with machine learning technology to automate land-use classifications and maintain consistent methods for classifying different types of land use.
- ☐ Incorporate imagery, including satellite, aerial, and drone images, to monitor and classify land use and land changes over time.
- ☐ Move away from outdated land management systems, such as paper or textual systems, and combine the data within GIS to better align land management decisions across agencies, departments, developers, and the public.
- ☐ Provide land managers and users with spatial tools to edit, analyze, and share land-use data within a consistent digital framework.
- ☐ Use GIS to generate and evaluate regular inventories of different types of land use, such as parks, wilderness, wild and scenic rivers, and other protected areas, and prioritize actions needed to support other landowning organizations, including nongovernmental organizations.

Climate and conservation

The nature of land changes over time because of development, climate change, political upheaval, and declining ecosystems. Often, these changes affect the economic and social well-being of communities, cause wildlife habitat loss, and reduce the viability of sustainable land practices. Conservationists can use GIS to track wildlife conflicts and engage stakeholders in and around protected areas, including monitoring protection operations and visualizing patterns of illegal activities, such as poaching. As the real-life examples in this book showed, the use of geospatial tools to respond to climate change risks and ecosystem threats helps organizations better anticipate and remedy the impacts to communities and operations. Spatial data and GIS can help identify ecological concerns and define a better path to sustainable prosperity.

Climate and conservation checklist

- ☐ Preserve habitat and open spaces by identifying valuable parcels and land areas using to GIS to visualize, analyze, and organize spatial data, imagery, and evaluations.
- ☐ Use GIS maps and applications to educate landowners, growing local support and increasing community involvement in efforts to preserve habitat.
- ☐ Prioritize land for conservation, using GIS to accurately identify, describe, and prioritize land parcels and areas for conservation.
- ☐ Analyze agricultural land and proposed development areas in conjunction with vulnerable ecosystems to decrease the potential for habitat fragmentation.

- Respond to flooding and shoreline loss because of climate change by mapping and spatially quantifying the impact to watersheds, creeks, and shorelines.
- Use GIS mapping and analysis tools to understand and continuously measure the chemical, physical, and biological health of wetlands, estuaries, and river systems.
- Deploy mobile workers with GIS mobile applications to collect field data, photos, and videos to update assessments of land quality.
- Investigate the costs and benefits of sustainable development needed to preserve biodiversity, protect ecosystems, and reduce poverty.
- Share conservation maps with partner organizations and work together to formulate strategies to regulate land use, apply for grants, and apply for tax credits.

Equity and social justice

For many people, land and property ownership are precious resources. Unfortunately, the value of land as a commodity and for its cultural significance can be mishandled and corrupted, leading to conflict and the displacement of people. With GIS, land administration agencies and nonprofit organizations can use location intelligence to better understand the commercial and social costs of land inequity. GIS can help users find new ways to address the issues of impoverished communities and empower displaced people by applying scientific tools to gain a stronger voice.

Advancements in geospatial data analysis and mapping help land administrators address the policies and practices that negatively impact disadvantaged people and see how those actions reflect on the whole community. GIS can help mitigate economic and environmental disruption, diminution of cultural heritage, and unethical management of land and property.

Equity and social justice checklist

- ☐ Empower women landowners by using GIS maps, analysis, and apps to digitize land records. View property parcels on a map, and legitimize land use, ownership, and land value.
- ☐ Overcome land dispossession by mapping GIS historic information about property abandonment and deprivation, giving displaced people and marginalized cultures better access to information about their land rights.
- ☐ Use GIS maps and mobile data collection applications to document landownership and measure the impacts of poverty, evictions, safety, and health of low-income and displaced people.
- ☐ Plan for affordable housing by using GIS to visualize and better understand affordable housing opportunities, and ensure diversity in income, race, and other demographic factors.

How to get started with GIS

You can apply GIS to land administration in two ways: Esri solutions and hands-on learning.

Esri solutions for land administration are designed for quickly applying GIS maps and apps for these purposes:

- **Development impact analysis:** Create effective 3D visualizations that highlight the impact of proposed developments.
- **Parcel value analysis:** Use GIS to connect with assessment and tax business systems; narrow a search of property sales by defining a price range, year built, and living area; and provide a high-level view into changes in property assessments and tax revenue within a local government.
- **Tax reverted property:** Improve collection of delinquent property information, document tax-delinquent properties, and market properties during a property tax sale.
- **Address data management:** Maintain a standard set of address layers to incorporate national guidelines, streamline address data management workflows, and continuously improve the quality of address information.
- **Public notification:** Identify properties, generate a list of owners or occupants, and document which properties were included in the public notice.
- **Conservation easement monitoring:** Understand current property conditions, inspect conservation easements, monitor conservation easement programs, and engage conservation easement stakeholders.

- **Environmental analysis:** Evaluate proposed development projects and solicit stakeholder feedback.
- **Conservation outreach:** Capture conflicts that arise from interactions between humans and wildlife, manage the resolution of each conflict, and monitor outreach activities in and around protected areas.
- **Equitable property value:** Communicate parcel, property sales, and floodplain information, and solicit assessment appeals from property owners to improve customer service and promote fair and equitable property values.

Hands-on learning will increase your understanding of GIS and help you use the technology to improve land data integrity and overall usefulness. A good place to start is Learn ArcGIS, an online collection of free story-driven lessons that illustrate how GIS applies to real-life problems. Learn ArcGIS lessons can support land administration practices in these ways:

- **Work with ArcGIS Parcel Fabric:** Learn ArcGIS lessons and videos provide various training about using Parcel Fabric, including working with legal descriptions of property, tracking the history and lineage of changes to the land record, dividing a parcel into equal areas, and using legal parcel records to update the parcel fabric.
- **Evaluate locations for mixed-use development:** Research suitably zoned areas populated with young adults near light-rail stops, and then share your knowledge with your organization.

- **Plan scenarios for future land use:** Learn how to work with a general plan's land use element using a 3D urban model.
- **Explore commercial permits:** Analyze permit data to detect patterns and understand why growth is occurring, and then share your work online.
- **Measure and visualize digital exclusion:** Better understand the scale and breadth of racial disparities and their causal factors, which in turn can lead to more effective solutions.
- **Calculate environmental equity for public policy:** Examine toxic chemical release exposure risk to different population subgroups from facilities in the federal Environmental Protection Agency's Toxics Release Inventory and assess environmental equity.

Learn about additional GIS resources for land administration by visiting the web page for this book:

go.esri.com/vpap-resources

CONTRIBUTORS

Jordan Baltierra, City of Newport Beach, California
Anne Cassidy, Winrock International
Mason Croft, Boise State University
David Gadsden, Esri
Mark Henry, Esri
Angela Hronek, Mead & Hunt
Lisa Johnson, Boise State University
Meghan T. Jones, LJB Inc.
Ilyanna Kadich, Texas Water Development Board
David Oberman, Esri
Larry Orman, GreenInfo Network
Monica Pratt, Esri
Sebastian Renfield, Mead & Hunt
Victoria Roberts, Esri
Citabria Stevens, Esri
Jay Strahan, Los Angeles Department of City Planning

ABOUT ESRI PRESS

AT ESRI PRESS, OUR MISSION IS TO INFORM, INSPIRE, AND teach professionals, students, educators, and the public about GIS by developing print and digital publications. Our goal is to increase the adoption of ArcGIS and to support the vision and brand of Esri. We strive to be the leader in publishing great GIS books, and we are dedicated to improving the work and lives of our global community of users, authors, and colleagues.

Acquisitions

Stacy Krieg
Claudia Naber
Alycia Tornetta
Craig Carpenter
Jenefer Shute

Editorial

Carolyn Schatz
Mark Henry
David Oberman

Production

Monica McGregor
Victoria Roberts

Marketing

Mike Livingston
Sasha Gallardo
Beth Bauler

Contributors

Christian Harder
Matt Artz
Keith Mann

Business

Catherine Ortiz
Jon Carter
Jason Childs

For information on Esri Press books and resources, visit our website at esri.com/en-us/esri-press.